Practice Management (PcM)

ARE 5 Mock Exam
(Architect Registration Examination)

ARE 5 Overview, Exam Prep Tips,
Hotspots, Case Studies, Drag-and-Place,
Solutions and Explanations

Gang Chen

ArchiteG®, Inc.
Irvine, California

Practice Management (PcM) ARE 5.0 Mock Exam (Architect Registration Examination): ARE 5.0 Overview, Exam Prep Tips, Hotspots, Case Studies, Drag-and-Place, Solutions and Explanations

Copyright © 2018   Gang Chen
V7
Cover Photo © 2018   Gang Chen

Copy Editor: Penny L Kortje

All Rights Reserved.
No part of this book may be transmitted or reproduced by any means or in any form, including electronic, graphic, or mechanical, without the express written consent of the publisher or author, except in the case of brief quotations in a review.

ArchiteG®, Inc.
http://www.ArchiteG.com

ISBN: 9781612650388

PRINTED IN THE UNITED STATES OF AMERICA

What others are saying about *ARE Mock Exam series* ...

"Great study guide..."
"This was a great resource supplement to my other study resources. I appreciated the mock exam questions the most, and the solutions offer an explanation as to why the answer is correct. I will definitely check out his other ARE exam resources!

UPDATE: Got my PASS Letter!"
—Sean Primeaux

"Tried everything 4 times before reading this book and PASSED!"
"I had failed this exam 4 times prior to getting this book…I had zero clue as to what I was doing wrong. I read Ballast, Kaplan and random things on the forum but for the life of me couldn't pin point where I was missing it until I read THIS BOOK! Gang did an excellent job…I remember … reading Gang's book and saying Ohhhh like 4 or 5 times. I read his book several times until I became comfortable with the information. I went in on test day and it was a breeze. I remember walking out of there thinking I couldn't believe I struggled so much before. The tips in here are priceless! I strongly recommend this book…"
—hendea1

"Add this to your ARE study"
"This was a very helpful practice exam and discussion. I really appreciated the step-by-step review of the author's approach... As I studied it last before taking the test, Gang Chen's book probably made the difference for me."
—Dan Clowes ("XLine")

"Good supplemental mock exam"
"I found the mock exam to be very helpful, all of the answers are explained thoroughly and really help you understand why it is correct...Also the introduction and test taking tips are very helpful for new candidates just starting the ARE process."
—Bgrueb01

"Essential Study Tool"
"I have read the book and found it to be a great study guide for myself. Mr. Gang Chen does such a great job of helping you get into the right frame of mind for the content of the exam. Mr. Chen breaks down the points on what should be studied and how to improve your chances of a pass with his knowledge and tips for the exam.

I highly recommend this book to anyone…it is an invaluable tool in the preparation for the exam as Mr. Chen provides a vast amount of knowledge in a very clear, concise, and logical matter."
—Luke Giaccio

"Wish I had this book earlier"
"...The questions are written like the NCARB questions, with various types...check all that apply, fill in the blank, best answer, etc. The answer key helpfully describes why the correct answer is correct, and why the incorrect answers are not. Take it from my experience, at half the cost of other mock exams, this is a must buy if you want to pass..."
—**Domiane Forte ("Vitruvian Duck")**

"This book did exactly like the others said."
"This book did exactly like the others said. It is immensely helpful with the explanation... There are so many codes to incorporate, but Chen simplifies it into a methodical process. Bought it and just found out I passed. I would recommend."
—**Dustin**

"It was the reason I passed."
"This book was a huge help. I passed the AREs recently and I felt this book gave me really good explanations for each answer. It was the reason I passed."
—**Amazon Customer**

"Great Practice Exam"
"… For me, it was difficult to not be overwhelmed by the amount of content covered by the Exam. This Mock Exam is the perfect tool to keep you focused on the content that matters and to evaluate what you know and what you need to study. It definitely helped me pass the exam!!"
—**Michael Harvey ("Harv")**

"One of the best practice exams"
"Excellent study guide with study tips, general test info, and recommended study resources. Hands down one of the best practice exams that I have come across for this exam. Most importantly, the practice exam includes in depth explanations of answers. Definitely recommended."
—**Taylor Cupp**

"Great Supplement!!"
"This publication was very helpful in my preparation for my BS exam. It contained a mock exam, followed by the answers and brief explanations to the answers. I would recommend this as an additional study material for this exam."
—**Cynthia Zorrilla-Canteros ("czcante")**

"Fantastic! "
"When I first began to prepare for this exam; the number of content areas seemed overwhelming and daunting at best. However, this guide clearly dissected each content area into small management components. Of all the study guides currently available for this test - this exam not only included numerous resources (web links, you tube clips, etc..), but also the sample test was extremely helpful. The sample test incorporated a nice balance of diagrams, calculations and general concepts - this book allowed me to highlight any "weak" content areas I had prior to the real exam. In short - this is an awesome book!"
—**Rachel Casey (RC)**

Dedication

To my parents, Zhuixian and Yugen,
and my daughters,
Alice, Angela, Amy, and Athena.

Disclaimer

Practice Management (PcM) ARE 5.0 Mock Exam (Architect Registration Examination) provides general information about Architect Registration Examination. The book is sold with the understanding that neither the publisher nor the authors are providing legal, accounting, or other professional services. If legal, accounting, or other professional services are required, seek the assistance of a competent professional firm.

The purpose of this publication is not to reprint the content of all other available texts on the subject. You are urged to read other materials, and tailor them to fit your needs.

Great effort has been taken to make this resource as complete and accurate as possible. However, nobody is perfect and there may be typographical errors or other mistakes present. You should use this book as a general guide and not as the ultimate source on this subject. If you find any potential errors, please send an e-mail to:
info@ArchiteG.com

Practice Management (PcM) ARE 5.0 Mock Exam (Architect Registration Examination) is intended to provide general, entertaining, informative, educational, and enlightening content. Neither the publisher nor the author shall be liable to anyone or any entity for any loss or damages, or alleged loss or damages, caused directly or indirectly by the content of this book.

ArchiteG®, Green Associate Exam Guide®, GA Study®, and GreenExamEducation® are registered trademarks owned by Gang Chen.

ARE®, Architect Registration Examination® are registered trademarks owned by NCARB.

If you do not wish to be bound by the above, you may return this book to the publisher for a full refund.

Legal Notice

ARE Mock Exam series by ArchiteG, Inc.

Time and effort is the most valuable asset of a candidate. How to cherish and effectively use your limited time and effort is the key of passing any exam. That is why we publish the ARE Mock Exam series to help you to study and pass the ARE exams in the shortest time possible. We have done the hard work so that you can save time and money. We do not want to make you work harder than you have to. To save your time, we use a *standard* format for all our ARE 5.0 Mock Exam books, so that you can quickly skip the *identical* information you have already read in other books of the series and go straight to the *unique* "meat and potatoes" portion of the book.

The trick and the most difficult part of writing a good book is to turn something that is very complicated into something that is very simple. This involves researching and really understanding some very complicated materials, absorbing the information, and then writing about the topic in a way that makes it very easy to understand. To succeed at this, you need to know the materials very well. Our goal is to write books that are clear, concise, and helpful to anyone with a seventh-grade education.

Do not force yourself to memorize a lot of numbers. Read through the numbers a few times, and you should have a very good impression of them.

You need to make the judgment call: If you miss a few numbers, you can still pass the exam, but if you spend too much time drilling these numbers, you may miss out on the big pictures and fail the exam.

The existing ARE practice questions or exams by others are either way too easy or way over-killed. They do NOT match the real ARE 5.0 exams at all.

We have done very comprehensive research on the official NCARB guides, many related websites, reference materials, and other available ARE exam prep materials. We match our mock exams as close as possible to the NCARB samples and the real ARE exams. Some readers had failed an ARE exam two or three times before, and they eventually passed the exam with our help.

All our books include a complete set of questions and case studies. We try to mimic the real ARE exams by including the same number of questions, using a similar format, and asking the same type of questions. We also include detailed answers and explanations to our questions.

There is some extra information on ARE overviews and exam-taking tips in Chapter One. This is based on NCARB *and* other valuable sources. This is a bonus feature we included in each book because we want our readers to be able to buy our ARE mock exam books together or individually. We want you to find all necessary ARE exam information and resources at one place and through our books.

All our books are available at
http://www.GreenExamEducation.com

How to Use This Book

We suggest you read *Practice Management (PcM) ARE 5.0 Mock Exam (Architect Registration Examination)* at least three times:

Read once and cover Chapter One and Two, the Appendixes, the related *free* PDF files, and other resources. Highlight the information you are not familiar with.

Read twice focusing on the highlighted information to memorize. You can repeat this process as many times as you want until you master the content of the book.

After reviewing these materials, you can take the mock exam, and then check your answers against the answers and explanations in the back, including explanations for the questions you answer correctly. You may have answered some questions correctly for the wrong reason. Highlight the information you are not familiar with.

Like the real exam, the mock exam will continue to use **multiple choice, check-all-that-apply,** and **quantitative fill-in-the-blank**. There are also three new question types: **Hotspots, case studies,** and **drag-and-place**.

Review your highlighted information and take the mock exam again. Try to answer 100% of the questions correctly this time. Repeat the process until you can answer all the questions correctly.

This book includes most if not all the information you need to do the calculations, as well as step-by-step explanations. After reading this book, you will greatly improve your ability to deal with the real ARE questions and have a great chance of passing the exam on the first try.

Take the mock exam at least two weeks before the real exam. You should *not* wait until the night before the real exam to take the mock exam. If you do not do well, you will go into panic mode and NOT have enough time to review your weaknesses.

Read for the final time the night before the real exam. Review *only* the information you highlighted, especially the questions you did not answer correctly when you took the mock exam for the first time.

This book is very light so you can easily carry it around. These features will allow you to review the book whenever you have a few minutes.

The Table of Contents is very detailed so you can locate information quickly. If you are on a tight schedule you can forgo reading the book linearly and jump around to the sections you need.

All our books, including "ARE Mock Exams Series" and "LEED Exam Guides Series," are available at
GreenExamEducation.com

Check out FREE tips and info at **GeeForum.com**, you can post your questions for other users' review and responses.

Table of Contents

Dedication..5

Disclaimer..7

ARE Mock Exam Series by ArchiteG, Inc..9

How to Use This Book..11

Table of Contents..13

Chapter One   Overview of Architect Registration Examination (ARE)

    A. **First Thing First: Go to the Website of your Architect Registration Board and Read all the Requirements of Obtaining an Architect License in your Jurisdiction**..17

    B. **Download and Review the Latest ARE Documents at the NCARB Website**..17
       1. Important links to the FREE and official NCARB documents
       2. A detailed list and brief description of the FREE PDF files that you can download from NCARB
          - ARE 5.0 Credit Model
          - ARE 5.0 Guidelines
          - NCARB Education Guidelines
          - Architectural Experience Program (AXP) Guidelines
          - Certification Guidelines
          - ARE 5.0 Related FAQs (Frequently Asked Questions)
          - Your Guide to ARE 5.0
          - ARE 5.0 Handbook
          - ARE 5.0 Test Specification
          - ARE 5.0 Prep Videos
          - The Burning Question: Why Do We need an ARE anyway?
          - Defining Your Moral Compass
          - Rules of Conduct

    C. **The Intern Development Program (IDP)/Architectural Experience Program (AXP)**..21
       1. What is IDP? What is AXP?
       2. Who qualifies as an intern?

D. Overview of Architect Registration Examination (ARE)..22
1. How to qualify for the ARE?
2. How to qualify for an architect license?
3. What is the purpose of ARE?
4. What is NCARB's rolling clock?
5. How to register for an ARE exam?
6. How early do I need to arrive at the test center?
7. Exam Format & Time
    - Practice Management (PcM)
    - Project Management (PjM)
    - Programming & Analysis (PA)
    - Project Planning & Design (PPD)
    - Project Development & Documentation (PDD)
    - Construction & Evaluation (CE)
8. How are ARE scores reported?
9. Is there a fixed percentage of candidates who pass the ARE exams?
10. When can I retake a failed ARE division?
11. How much time do I need to prepare for each ARE division?
12. Which ARE division should I take first?
13. ARE exam prep and test-taking tips
14. Strategies for passing ARE exams on the first try
    - Find out how much you already know and what you should study
    - Cherish and effectively use your limited time and effort
    - Do NOT stretch your exam prep process too long
    - Resist the temptation to read too many books and limit your time and effort to read only a few selected books or a few sections of books in details
    - Think like an architect
15. ARE exam preparation requires short-term memory
16. Allocation of your time and scheduling
17. Timing of review: the 3016 rule; memorization methods, tips, suggestions, and mnemonics
18. The importance of good and effective study methods
19. The importance of repetition: read this book <u>at least</u> three times
20. The importance of a routine
21. The importance of short, frequent breaks and physical exercise
22. A strong vision and a clear goal
23. Codes and standards used in this book
24. Where can I find study materials on architectural history?

Chapter Two   Practice Management (PcM) Division

A. General Information..37
1. Exam content
2. Official exam guide and reference index for the Practice Management (PcM) division

B. **The Most Important Documents/Publications for PcM Division of the ARE Exam.38**
   1. *Official NCARB list of references for the Practice Management (PcM) division with our comments and suggestions*
      Publications; AIA Contract Documents
   2. Construction Specifications Institute (CSI) MasterFormat & *Building Construction*

Chapter Three   ARE Mock Exam for Practice Management (PcM) Division

A. Multiple-Choice (MC) ……….............…………………………….....………………....43

B. Case Study………........…………………………………………….…………………58

Chapter Four   ARE Mock Exam Solutions for Practice Management (PcM) Division

A. Mock Exam Answers and Explanations: Multiple-Choice (MC)…………..................63

B. Mock Exam Answers and Explanations: Case Study...………....…………………83

Appendixes
   A. List of Figures...……………………………………………….…….....................89
   B. Official reference materials suggested by NCARB…….....……………….................90
      1. Resources Available While Testing
      2. Typical Beam Nomenclature
      3. Formulas Available While Testing
      4. Common Abbreviations
      5. General NCARB reference materials for ARE
      6. Official NCARB reference materials matrix
      7. Extra Study Materials
   C. Other reference materials…………………….....……………...…….....…...106
   D. Some Important Information about Architects and the Profession of Architecture…………………………..……………………………………107
   E. AIA Compensation Survey…………………….....……………...………….113
   F. So … You would Like to Study Architecture…………....…...……...………114

Back Page Promotion
   A. ARE Mock Exam series (GreenExamEducation.com)
   B. LEED Exam Guides series (GreenExamEducation.com)
   C. *Building Construction* (ArchiteG.com)
   D. *Planting Design Illustrated*

Index

Chapter One

Overview of the Architect Registration Examination (ARE)

A. **First Thing First: Go to the Website of your Architect Registration Board and Read all the Requirements of Obtaining an Architect License in your Jurisdiction**
See the following link:
https://www.ncarb.org/get-licensed/state-licensing-boards

B. **Download and Review the Latest ARE Documents at the NCARB Website**

1. **Important links to the FREE and official NCARB documents**
NCARB launched ARE 5.0 on November 1, 2016. ARE 4.0 will continue to be available until June 30, 2018.

   ARE candidates who started testing in ARE 4.0 can choose to "self-transition" to ARE 5.0. This will allow them to continue testing in the version that is most suitable for them. However, **once a candidate transitions to ARE 5.0, s/he cannot transition back to ARE 4.0**.

   The current version of the Architect Registration Examination (ARE 5.0) includes six divisions:

   - Practice Management (PcM)
   - Project Management (PjM)
   - Programming & Analysis (PA)
   - Project Planning & Design (PPD)
   - Project Development & Documentation (PDD)
   - Construction & Evaluation (CE)

   All ARE divisions continue to use **multiple choice, check-all-that-apply,** and **quantitative fill-in-the-blank**. The new exams include three new question types: **Hotspots, case studies,** and **drag-and-place**.

   There is a tremendous amount of valuable information covering every step of becoming an architect available free of charge at the NCARB website:
   http://www.ncarb.org/

   For example, you can find guidance about architectural degree programs accredited by the National Architectural Accrediting Board (NAAB), NCARB's Architectural Experience Program (AXP) formerly known as Intern Development Program (IDP), and licensing requirements by state. These documents explain how you can qualify to take the Architect Registration Examination.

We find the official ARE 5.0 Guidelines, ARE 5.0 Handbook, and ARE 5.0 Credit Model extremely valuable. See the following link:
http://www.ncarb.org/ARE/ARE5.aspx

You should start by studying these documents.

2. **A detailed list and brief description of the FREE PDF files that you can download from NCARB**
The following is a detailed list of the FREE PDF files that you can download from NCARB. They are listed in order based on their importance.

- All **ARE 5.0** information can be found at the following links:
  http://www.ncarb.org/ARE/ARE5.aspx
  http://blog.ncarb.org/2016/November/ARE5-Study-Materials.aspx
- The **ARE 5.0 Credit Model** is one of the most important documents and tells you the relationship between ARE 4.0 and ARE 5.0.

ARE5.0:	Practice Management	Project Management	Programming & Analysis	Project Planning & Design	Project Development & Documentation	Construction & Evaluation
ARE 4.0:						
Construction Documents & Services	●	●			●	●
Programming Planning & Practice	●	●	●			
Site Planning & Design			●	●		
Building Design & Construction Systems				●	●	
Structural Systems				●	●	
Building Systems				●	●	
Schematic Design				●		

Figure 1.1 The relationship between ARE 4.0 and ARE 5.0

- **ARE 5.0 Guidelines** includes extremely valuable information on the ARE overview, NCARB, registration (licensure), architectural education requirements, the Architectural Experience Program (AXP), establishing your eligibility to test, scheduling an exam appointment, taking the ARE, receiving your score, retaking the ARE, the exam format, scheduling, and links to other FREE NCARB PDF files. You need to read this <u>at least twice</u>.

- **NCARB Education Guidelines** contains information on education requirements for initial licensure and for NCARB certification, satisfying the education requirement, foreign-educated applicants, the education alternative to NCARB certification, the Education Evaluation Services for Architects (EESA), the Education Standard, and other resources.

- **Architectural Experience Program (AXP) Guidelines** includes information on AXP overview, getting started and creating your NCARB record, experience areas and tasks, documenting your experience through hours, documenting your experience through a portfolio, and the next steps. You need to read this document <u>at least twice</u>. The information is valuable.

NCARB renamed the **Intern Development Program (IDP)** as **Architectural Experience Program (AXP)** in June 2016. Most of NCARB's 54-member boards have adopted the AXP as a prerequisite for initial architect licensure. Therefore, you should be familiar with the program.

The AXP application fee is $100. This fee includes one free transmittal of your Record for initial registration and keeps your Record active for the first year. After the initial year, there is an annual renewal fee required to maintain an active Record until you become registered. The cost is currently $85 each year. The fees are subject to change, and you need to check the NCARB website for the latest information.

There are two ways to meet the AXP requirements. The **first method** is **reporting hours**. Most candidates will use this method. You will need to document at least 3,740 required hours under the six different experience areas to complete the program. A minimum of 50% of your experience must be completed under the supervision of a qualified architect.

The following chart lists the hours required under the six experience areas:

Experience Area	Hours Required
Practice Management	160
Project Management	360
Programming & Analysis	260
Project Planning & Design	1,080
Project Development & Documentation	1,520
Construction & Evaluation	360
Total	**3,740**

Figure 1.2 The hours required under the six experience areas

Your experience reports will fall under one of **two experience settings**:
- **Setting A**: Work performed for an architecture firm.
- **Setting O**: Experiences performed outside an architecture firm.

You must earn at least **1,860 hours** in experience **setting A**.

Your AXP experience should be reported to NCARB at least every six months and logged within two months of completing each reporting period (the **Six-Month Rule**).

The **second method** to meet AXP requirements is to create an **AXP Portfolio**. This new method is for experienced design professionals who put their licensure on hold and allows you to prove your experience through the preparation of an online portfolio.

To complete the AXP through the **second method**, you will need to meet ALL the AXP criteria through the portfolio. In other words, you cannot complete the experience requirement through a combination of the **AXP portfolio** and **reporting hours**.

See the following link for more information on AXP:
https://www.ncarb.org/gain-axp-experience

- **Certification Guidelines** by NCARB (Skimming through this should be adequate. You should also forward a copy of this PDF file to your AXP supervisor.)

  See the following link which contains resources for supervisors and mentors:
  http://www.ncarb.org/Experience-Through-Internships/Supervisors-and-Mentors/Resources-for-Supervisor-and-Mentors.aspx

- **ARE 5.0 Related FAQs (Frequently Asked Questions)**: Skimming through this should be adequate.

- **Your Guide to ARE 5.0** includes information on understanding the basics of ARE 5.0, new question types, taking the test, making the transition, getting ARE 5.0 done, and planning your budget. The document also contains FAQs, and links for more information. You need to read this document at least twice. The information is valuable.

- **ARE 5.0 Handbook** contains an ARE overview, detailed information for each ARE division, and ARE 5.0 references. This handbook explains what NCARB expects you to know so that you can pass the ARE exams. ARE 5.0 uses either **Understand/Apply (U/A)** or **Analyze/Evaluate (A/E)** to designate the appropriate cognitive complexity of each objective, but *avoids* the use of **"Remember,"** the lowest level of cognitive complexity (CC), or **"Create,"** the highest level of CC.

  This handbook has some sample questions for each division. The real exam is like the samples in this handbook.

Tips:
- *ARE 5.0 Handbook has about 180 pages. To save time, you can just read the generic information at the front and back portion of the handbook, and focus on the ARE division(s) you are currently studying for. As you progress in your testing, you can read the applicable division that you are studying for. This way, the content will always be fresh in your mind.*
- *You need to read this document <u>at least three times</u>. The information is valuable.*

- **ARE 5.0 Test Specification** identifies the ARE 5.0 division structure and defines the major content areas, called **Sections**; the measurement **Objectives**; and the percentage of content coverage, called **Weightings**. This document specifies the scope and objectives of each ARE division, and the percentage of questions in each content category. You need to read this document <u>at least twice</u>. The information is valuable, and is the base of all ARE exam questions.

- **ARE 5.0 Prep Videos** include one short video for each division. These videos give you a very good basic introduction to each division, including sample questions and answers, and explanations. You need to watch each video <u>at least three times</u>. See the following link:
  http://blog.ncarb.org/2016/November/ARE5-Study-Materials.aspx

- **The Burning Question: Why Do We Need an ARE Anyway?** (We do not want to give out a link for this document because it is too long and keeps changing. You can Google it with its title. Skimming through this document should be adequate.)

- **Defining Your Moral Compass** (You can Google it with its title plus the word "NCARB." Skimming through this document should be adequate.)

- **Rules of Conduct** is available as a FREE PDF file at:
  http://www.ncarb.org/
  (Skimming through this should be adequate.)

C. The Intern Development Program (IDP)/Architectural Experience Program (AXP)

1. What is IDP? What is AXP?
IDP is a comprehensive training program jointly developed by the National Council of Architectural Registration Boards (NCARB) and the American Institute of Architects (AIA) to ensure that interns obtain the necessary skills and knowledge to practice architecture <u>independently</u>. NCARB renamed the **Intern Development Program (IDP)** as **Architectural Experience Program (AXP)** in June 2016.

2. Who qualifies as an intern?
Per NCARB, if an individual meets one of the following criteria, s/he qualifies as an intern:
a. Graduates from NAAB-accredited programs
b. Architecture students who acquire acceptable training prior to graduation
c. Other qualified individuals identified by a registration board

D. Overview of the Architect Registration Examination (ARE)

1. **How to qualify for the ARE?**
   A candidate needs to qualify for the ARE via one of NCARB's member registration boards, or one of the Canadian provincial architectural associations.

   Check with your Board of Architecture for specific requirements.

   For example, in California, a candidate must provide verification of a minimum of <u>five</u> years of education and/or architectural work experience to qualify for the ARE.

   Candidates can satisfy the five-year requirement in a variety of ways:

   - Provide verification of a professional degree in architecture through a program that is accredited by NAAB or CACB.

     OR
   - Provide verification of at least five years of educational equivalents.

     OR
   - Provide proof of work experience under the direct supervision of a licensed architect.

   See the following link:
   http://www.ncarb.org/ARE/Getting-Started-With-the-ARE/Ready-to-Take-the-ARE-Early.aspx

2. **How to qualify for an architect license?**
   Again, each jurisdiction has its own requirements. An individual typically needs a combination of about <u>eight</u> years of education and experience, as well as passing scores on the ARE exams. See the following link:
   http://www.ncarb.org/Reg-Board-Requirements

   For example, the requirements to become a licensed architect in California are:
   - Eight years of post-secondary education and/or work experience as evaluated by the Board (including at least one year of work experience under the direct supervision of an architect licensed in a U.S. jurisdiction or two years of work experience under the direct supervision of an architect registered in a Canadian province)
   - Completion of the Architectural Experience Program (AXP)
   - Successful completion of the Architect Registration Examination (ARE)
   - Successful completion of the California Supplemental Examination (CSE)

   California does NOT require an accredited degree in architecture for examination and licensure. However, many other states do.

3. **What is the purpose of ARE?**
   The purpose of ARE is NOT to test a candidate's competency on every aspect of architectural practice. Its purpose is to test a candidate's competency on providing professional services to protect the <u>health, safety, and welfare</u> of the public. It tests candidates on the <u>fundamental</u> knowledge of pre-design, site design, building design, building systems, and construction documents and services.

   The ARE tests a candidate's competency as a "specialist" on architectural subjects. It also tests her abilities as a "generalist" to coordinate other consultants' works.

   You can download the exam content and references for each of the ARE divisions at the following link:
   https://www.ncarb.org/pass-the-are/start

4. **What is NCARB's rolling clock?**
   a. Starting on January 1, 2006, a candidate MUST pass ALL ARE sections within five years. A passing score for an ARE division is only valid for five years, and a candidate has to retake this division if she has NOT passed all divisions within the five-year period.

   b. Starting on January 1, 2011, a candidate who is authorized to take ARE exams MUST take at least one division of the ARE exams within five years of the authorization. Otherwise, the candidate MUST apply for the authorization to take ARE exams from an NCARB member board again.

   These rules were created by the **NCARB's rolling clock** resolution and passed by NCARB council during the 2004 NCARB Annual Meeting.

   ARE 4.0 division expiration dates per the Rolling Clock will remain the same for the transition to ARE 5.0.

5. **How to register for an ARE exam?**
   See the instructions in the new ARE guideline at the following link:
   http://www.ncarb.org/ARE/ARE5.aspx

6. **How early do I need to arrive at the test center?**
   Be at the test center at least 30 minutes BEFORE your scheduled test time, OR you may lose your exam fee.

7. **Exam format & time**
All ARE divisions are administered and graded by computer. The time allowances are as follows:

Division	Number of Questions	Test Duration	Appointment Time
Practice Management	80	2 hr 45 min	3 hr 30 min
Project Management	95	3 hr 15 min	4 hr
Programming & Analysis	95	3 hr 15 min	4 hr
Project Planning & Design	120	4 hr 15 min	5 hr
Project Development & Documentation	120	4 hr 15 min	5 hr
Construction & Evaluation	95	3 hr 15 min	4 hr
Total Time:		21 hr	25 hr 30 min

Figure 1.3   Exam format & time

Remote proctoring will be introduced mid December 2020. After December 13, 2020, the number of questions and time allotted will change to accommodate remote proctoring:

Division	Number of Questions	Test Duration	Appointment Time
Practice Management	65	2 hr 40 min	3 hr 20 min
Project Management	75	3 hr	3 hr 40 min
Programming & Analysis	75	3 hr	3 hr 40 min
Project Planning & Design	100	4 hr 5 min	5 hr
Project Development & Documentation	100	4 hr 5 min	5 hr
Construction & Evaluation	75	3 hr	3 hr 40 min
Total Time:		19 hr 50 min	24 hr 20 min

Figure 1.4   New Exam format & time

NCARB suggests you arrive at the test center a minimum of 30 minutes before your scheduled appointment. You can have one flexible 15-minute break for each division. That is why the appointment time is 45 minutes longer than the actual test time for each division.

Practice Management has 80 questions and NCARB allows you 2 hours and 45 minutes to complete the exam, so you should spend an average of $(2 \times 60 + 45)/80 = 165/80 = 2.06$ minutes on each question.

Project Management and **Programming & Analysis** as well as **Construction & Evaluation** each have 95 questions and NCARB allows you 3 hours and 15 minutes to complete each exam, so you should spend an average of $(3 \times 60 + 15)/80 = 195/95 = 2.05$ minutes on each question.

Project Planning & Design as well as **Project Development & Documentation** each have 120 questions and NCARB allows you 4 hours and 15 minutes to complete each exam, so you should spend an average of (4x60+15)/120=255/120= 2.13 minutes on each question.

To simplify this, we suggest you spend about 2 minutes for each question in ALL divisions.

8. How are ARE scores reported?
All ARE scores are reported as Pass or Fail. ARE scores are typically posted within 7 to 10 days. See the instructions in the new ARE guideline at the following link:
http://www.ncarb.org/ARE/ARE5.aspx

9. Is there a fixed percentage of candidates who pass the ARE exams?
No, there is NOT a fixed percentage of passing or failing. If you meet the minimum competency required to practice as an architect, you pass. The passing scores are the same for all Boards of Architecture.

10. When can I retake a failed ARE division?
You can retake a failed division of the ARE 60 days after the previous attempt. You can only take the same ARE division three (3) times within any 12-month period.

11. How much time do I need to prepare for each ARE division?
Every person is different, but on average you need about 40 to 80 hours to prepare for each ARE division. You need to set a realistic study schedule and stick with it. Make sure you allow time for personal and recreational commitments. If you are working full time, my suggestion is that you allow no less than 2 weeks but NOT more than 2 months to prepare for each ARE division. You should NOT drag out the exam prep process too long and risk losing your momentum.

12. Which ARE division should I take first?
This is a matter of personal preference, and you should make the final decision.

Some people like to start with the easier divisions and pass them first. This way, they build more confidence as they study and pass each division.

Other people like to start with the more difficult divisions so that if they fail, they can keep busy studying and taking the other divisions while the clock is ticking. Before they know it, six months have passed, and they can reschedule if need be.

13. ARE exam prep and test-taking tips
You can start with Construction & Evaluation (CE) because it gives a limited scope, and you may want to study building regulations and architectural history (especially famous architects and buildings that set the trends at critical turning points) before you take other divisions.

Complete mock exams and practice questions, including those provided by NCARB's practice program and this book, to hone your skills.

Form study groups and learn the exam experience of other ARE candidates. The forum at our website is a helpful resource. See the following links:
http://GreenExamEducation.com/
http://GeeForum.com/

Take the ARE exams as soon as you become eligible, since you probably still remember portions of what you learned in architectural school, especially structural and architectural history. Do not make excuses for yourself and put off the exams.

The following test-taking tips may help you:
- Pace yourself properly. You should spend about two minutes for each question on average.
- Read the questions carefully and pay attention to words like *best, could, not, always, never, seldom, may, false, except*, etc.
- For questions that you are not sure of, eliminate the obvious wrong answer and then make an educated guess. Please note that if you do NOT answer the question, you automatically lose the point. If you guess, you at least have a chance of getting it right.
- If you have no idea what the correct answer is and cannot eliminate any obvious wrong answers, then do not waste too much time on the question and just guess. Try to use the same guess answer for all of the questions you have no idea about. For example, if you choose "d" as the guess answer, then you should be consistent and use "d" whenever you have no clue. This way, you are likely have a better chance at guessing more answers correctly.
- Mark the difficult questions, answer them, and come back to review them AFTER you finish all questions. If you are still not sure, go with your first choice. Your first choice is often the best choice.
- You really need to spend time practicing to become VERY familiar with NCARB's question types. This is because ARE is a timed test, and you do NOT have time to learn about the question types during the test. If you do not know them well, you will NOT be able to finish your solution on time.
- The ARE exams test a candidate's competency to provide professional services protecting the health, safety, and welfare of the public. Do NOT waste time on aesthetic or other design elements not required by the program.

ARE exams are difficult, but if you study hard and prepare well, combined with your experience, AXP training, and/or college education, you should be able to pass all divisions and eventually be able to call yourself an architect.

14. Strategies for passing ARE exams on the first try

Passing ARE exams on the first try, like everything else, needs not only hard work, but also great strategy.

- **Find out how much you already know and what you should study**
  You goal is NOT to read all the study materials. Your goal is to pass the exam. Flip through the study materials. If you already know the information, skip these parts.

  Complete the NCARB sample questions for the ARE exam you are preparing for NOW without ANY studying. See what percentage you get right. If you get 68% right, you should be able to pass the real exam without any studying. If you get 50% right, then you just need 18% more to pass.

  This "truth-finding" exam or exercise will also help you to find out what your weakness areas are, and what to focus on.

  Look at the same questions again at the end of your exam prep, and check the differences.

  Note: We suggest you study the sample questions in the official NCARB Study Guide first, and then other study materials, and then come back to NCARB sample questions again several days before the real ARE exam.

  Per NCARB, with the launch of the updated Architect Registration Examination (ARE) 5.0 in December 2020, the new cutting scores are based on the following information:

  "**How Many Questions Do I Need Correct to Pass?**

  Each division of the ARE measures different content knowledge areas. The difference in knowledge areas and the relative difficulty of the questions that make up that content area vary between divisions; therefore, expectations around how many questions you will need to answer correctly also changes from division to division.

  - **Project Development & Documentation and Construction & Evaluation** require the lowest percentage of scored items to be answered correctly to pass. You need to answer between **57 – 62 percent** of scored items correctly on these divisions to pass.
  - **Practice Management and Project Management** require a slightly higher percentage of questions to be answered correctly to pass. You need to answer between **62 – 68 percent** of scored items correctly on these divisions to pass.
  - **Programming & Analysis and Project Planning & Design** require the highest percentage of questions to be answered correctly to pass. You need to answer between **65 – 71 percent** of scored items correctly on these divisions to pass."

  For detailed information, see the following link:
  https://www.ncarb.org/blog/what-score-do-you-need-to-pass-the-are

- **Cherish and effectively use your limited time and effort**
  Let me paraphrase a story.

One time someone had a chance to talk with Napoleon. He said:
"You are such a great leader and have won so many battles, that you can use one of your soldiers to defeat ten enemy soldiers."

Napoleon responded:
"That may be true, but I always try to create opportunities where ten of my soldiers fight one enemy soldier. That is why I have won so many battles."

Whether this story is true is irrelevant. The important thing that you need to know is **how to concentrate your limited time and effort to achieve your goal. Do NOT spread yourself too thin**. This is a principle many great leaders know and use and is why great leaders can use ordinary people to achieve extraordinary goals.

Time and effort is the most valuable asset of a candidate. How to cherish and effectively use your limited time and effort is the key to passing any exam.

If you study very hard and read many books, you are probably wasting your time. You are much better off picking one or two good books, covering the major framework of your exams, and then doing two sets of mock exams to find your weaknesses. You WILL pass if you follow this advice. You may still have minor weakness, but you will have covered your major bases.

- **Do NOT stretch your exam prep process too long**
  If you do this, it will hurt instead of helping you. You may forget the information by the time you take the exam.
  Spend 40 to 80 hours for each division (a maximum of two months for the most difficult exams if you really need more time) should be enough. Once you decide on taking an exam, put in 100% of your effort and read the RIGHT materials. Allocate your time and effort on the most important materials, and you will pass.

- **Resist the temptation to read too many books and limit your time and effort to read only a few selected books or a few sections of each book in detail**
  Having all the books but not reading them, or digesting ALL the information in them will not help you. It is like someone having a garage full of foods, and not eating or eating too much of them. Neither way will help.

You can only eat three meals a day. Similarly, you can ONLY absorb a certain amount of information during your exam prep. So, focus on the most important stuff.

Focus on your weaknesses but still read the other info. The key is to understand, digest the materials, and retain the information.

It is NOT how much you have read, but how much you understand, digest, and retain that counts.

The key to passing an ARE exam, or any other exam, is to know the scope of the exam, and not to read too many books. Select one or two really good books and focus on them, actually <u>understand</u> the content and <u>memorize</u> the information. For your convenience, I have <u>underlined</u> the fundamental information that I think is very important. You definitely need to <u>memorize</u> all the information that I have underlined.

You should try to understand the content first, and then memorize the content of the book by reading it multiple times. This is a much better way than relying on "mechanical" memory without understanding.

When you read the materials, ALWAYS keep the following in mind:

- **Think like an architect.**
  For example, when you take the Project Development & Documentation (PDD) exam, focus on what need to know to be able to coordinate your engineer's work, or tell them what to do. You are NOT taking an exam for becoming a structural engineer; you are taking an exam to become an architect.

  This criterion will help you filter out the materials that are irrelevant, and focus on the right information. You will know what to flip through, what to read line by line, and what to read multiple times.

  I have said it one thousand times, and I will say it once more:
  Time and effort is the most valuable asset of a candidate. How to cherish and effectively use your limited time and effort is the key to passing any exam.

15. ARE exam preparation requires short-term memory
You should understand that ARE Exam Preparation requires **Short-Term Memory**. This is especially true for the MC portion of the exam. You should schedule your time accordingly: in the <u>early</u> stages of your ARE exam Preparation, you should focus on <u>understanding</u> and an **<u>initial</u>** review of the material; in the <u>late</u> stages of your exam preparation, you should focus on <u>memorizing</u> the material as a **<u>final</u>** review.

16. Allocation of your time and scheduling
You should spend about 60% of your effort on the most important and fundamental study materials, about 30% of your effort on mock exams, and the remaining 10% on improving your weakest areas, i.e., reading and reviewing the questions that you answered incorrectly, reinforcing the portions that you have a hard time memorizing, etc.

Do NOT spend too much time looking for <u>obscure</u> ARE information because the NCARB will HAVE to test you on the most **<u>common</u>** architectural knowledge and information. At least <u>80% to 90%</u> of the exam content will have to be the most <u>common</u>, <u>important</u> and <u>fundamental</u> knowledge. The exam writers can word their questions to be <u>tricky</u> or <u>confusing</u>, but they have to limit themselves to the <u>important</u> content; otherwise, their tests will NOT be legally defensible. At most, <u>10%</u> of their test content can be <u>obscure</u> information. You only need to answer about 68% of all the questions. So, if you master the common ARE knowledge (applicable to 90% of the questions) and use the guess technique for the remaining 10% of the questions on the obscure ARE content, you will do well and pass the exam.

On the other hand, if you focus on the obscure ARE knowledge, you may answer the entire 10% <u>obscure</u> portion of the exam correctly, but only answer half of the remaining 90% of the <u>common</u> ARE knowledge questions correctly, and you will fail the exam. That is why we have seen many smart people who can answer very difficult ARE questions correctly because they are able to look them up and do quality research. However, they often end up failing ARE exams because they cannot memorize the common ARE knowledge needed on the day of the exam. ARE exams are NOT an open-book exams, and you cannot look up information during the exam.

The **process of memorization** is like **<u>filling a cup with a hole at the bottom</u>**: You need to fill it <u>faster</u> than the water leaks out at the bottom, and you need to <u>constantly</u> fill it; otherwise, it will quickly be empty.

Once you memorize something, your brain has already started the process of forgetting it. It is natural. That is how we have enough space left in our brain to remember the really important things.

It is tough to fight against your brain's natural tendency to forget things. Acknowledging this truth and the fact that you can<u>not</u> memorize everything you read, you need to <u>focus</u> your limited time, energy and brainpower on the <u>most important</u> issues.

The biggest danger for most people is that they memorize the information in the early stages of their exam preparation, but forget it before or on the day of the exam and still THINK they remember them.

Most people fail the exam NOT because they cannot answer the few "advanced" questions on the exam, but because they have read the information but can <u>NOT</u> recall it on the day of the exam. They spend too much time preparing for the exam, drag the preparation process on too long, seek too much information, go to too many websites, do too many practice questions and too many mock exams (one or two sets of mock exams can be good for you), and **spread themselves too thin**. They end up **missing the most important information** of the exam, and they will fail.

The ARE Mock Exam series along with the tips and methodology in each of the books will help you find and improvement your weakness areas, MEMORIZE the most important aspects of the test to pass the exam ON THE FIRST TRY.

So, if you have a lot of time to prepare for the ARE exams, you should plan your effort accordingly. You want your ARE knowledge to peak at the time of the exam, not before or after.

For example, <u>if you have two months to prepare for a very difficult ARE exam</u>, you may want to spend the first month focused on <u>reading and understanding</u> all of the study materials you can find as your **initial** review. Also, during this first month, you can start memorizing after you understand the materials as long as you know you HAVE to review the materials again later to retain them. If you have memorized something once, it is easier to memorize it again later.

Next, you can spend two weeks focused on memorizing the material. You need to review the material at least three times. You can then spend one week on mock exams. The last week before the exam, focus on retaining your knowledge and reinforcing your weakest areas. Read the mistakes that you have made and think about how to avoid them during the real exam. Set aside a mock exam that you have not taken and take it seven days before test day. This will alert you to your weaknesses and provide direction for the remainder of your studies.

If you have one week to prepare for the exam, you can spend two days reading and understanding the study material, two days repeating and memorizing the material, two days on mock exams, and one day retaining the knowledge and enforcing your weakest areas.

The last one to two weeks before an exam is absolutely critical. You need to have the "do or die" mentality and be ready to study hard to pass the exam on your first try. That is how some people are able to pass an ARE exam with only one week of preparation.

17. Timing of review: the 3016 rule; memorization methods, tips, suggestions, and mnemonics

Another important strategy is to review the material in a timely manner. Some people say that the best time to review material is between 30 minutes and 16 hours (the **3016** rule) after you read it for the first time. So, if you review the material right after you read it for the first time, the review may not be helpful.

I have personally found this method extremely beneficial. The best way for me to memorize study materials is to review what I learned during the day again in the evening. This, of course, happens to fall within the timing range mentioned above.

Now that you know the **3016** rule, you may want to schedule your review accordingly. For example, you may want to read new study materials in the morning and afternoon, then after dinner do an initial review of what you learned during the day.

OR

If you are working full time, you can read new study materials in the evening or at night and then get up early the next morning to spend one or two hours on an initial review of what you learned the night before.

The initial review and memorization will make your final review and memorization much easier.

Mnemonics is a very good way for you to memorize facts and data that are otherwise very hard to memorize. It is often arbitrary or illogical, but it works.

A good mnemonic can help you remember something for a long time or even a lifetime after reading it just once. Without the mnemonics, you may read the same thing many times and still not be able to memorize it.

There are a few common Mnemonics:
1) **Visual** Mnemonics: Link what you want to memorize to a visual image.
2) **Spatial** Mnemonics: link what you want to memorize to a space, and the order of things in it.
3) **Group** Mnemonics: Break up a difficult piece into several smaller and more manageable groups or sets, and memorize the sets and their order. One example is the grouping of the 10-digit phone number into three groups in the U.S. This makes the number much easier to memorize.
4) **Architectural** Mnemonics: A combination of Visual Mnemonics and Spatial Mnemonics and Group Mnemonics.

Imagine you are walking through a building several times, along the same path. You should be able to remember the order of each room. You can then break up the information that you want to remember and link them to several images, and then imagine you hang the images on walls of various rooms. You should be able to easily recall each group in an orderly manner by imagining you are walking through the building again on the same path, and looking at the images hanging on walls of each room. When you look at the images on the wall, you can easily recall the related information.

You can use your home, office or another building that you are familiar with to build an Architectural Mnemonics to help you to organize the things you need to memorize.

5) **Association** Mnemonics: You can associate what you want to memorize with a sentence, a similarly pronounced word, or a place you are familiar with, etc.
6) **Emotion** Mnemonics: Use emotion to fix an image in your memory.
7) **First Letter** Mnemonics: You can use the first letter of what you want to memorize to construct a sentence or acronym. For example, "**Roy G. Biv**" can be used to memorize the order of the 7 colors of the rainbow, it is composed of the first letter of each primary color.

You can use **Association** Mnemonics and memorize them as all the plumbing fixtures for a typical home, PLUS Urinal.

OR
You can use "Water S K U L" (**First Letter** Mnemonics selected from website below) to memorize them:

Water Closets
Shower
Kitchen Sinks
Urinal

Lavatory

18. The importance of good and effective study methods
There is a saying: Give a man a fish, feed him for a day. Teach a man to fish, feed him for a lifetime. I think there is some truth to this. Similarly, it is better to teach someone HOW to study than just give him good study materials. In this book, I give you good study materials to save you time, but more importantly, I want to teach you effective study methods so that you can not only study and pass ARE exams, but also so that you will benefit throughout the rest of your life for anything else you need to study or achieve. For example, I give you samples of mnemonics, but I also teach you the more important thing: HOW to make mnemonics.

Often in the same class, all the students study almost the SAME materials, but there are some students that always manage to stay at the top of the class and get good grades on exams. Why? One very important factor is they have good study methods.

Hard work is important, but it needs to be combined with effective study methods. I think people need to work hard AND work SMART to be successful at their work, career, or anything else they are pursuing.

19. The importance of repetition: read this book at least three times
Repetition is one of the most important tips for learning. That is why I have listed it under a separate title. For example, you should treat this book as part of the core study materials for your ARE exams and you need to read this book at least three times to get all of its benefits:

1) The first time you read it, it is new information. You should focus on understanding and digesting the materials, and also do an initial review with the **3016** rule.
2) The second time you read it, focus on reading the parts I have already highlighted AND you have highlighted (the important parts and the weakest parts for you).
3) The third time, focus on memorizing the information.

Remember the analogy of the memorization process as **filling a cup with a hole on the bottom**?
Do NOT stop reading this book until you pass the real exam.

20. The importance of a routine
A routine is very important for studying. You should try to set up a routine that works for you. First, look at how much time you have to prepare for the exam, and then adjust your current routine to include exam preparation. Once you set up the routine, stick with it.

For example, you can spend from 8:00 a.m. to 12:00 noon, and 1:00 p.m. to 5:00 p.m. on studying new materials, and 7:00 p.m. to 10:00 p.m. to do an initial review of what you learned during the daytime. Then, switch your study content to mock exams, memorization and retention when it gets close to the exam date. This way, you have 11 hours for exam preparation every day. You can probably pass an ARE exam in one week with this method. Just keep repeating it as a way to retain the architectural knowledge.

OR

You can spend 7:00 p.m. to 10:00 p.m. on studying new materials, and 6:00 a.m. to 7:00 a.m. to do an initial review of what you learned the evening before. This way, you have four hours for exam preparation every day. You can probably pass an ARE exam in two weeks with this preparation schedule.

A routine can help you to memorize important information because it makes it easier for you to concentrate and work with your body clock.

Do NOT become panicked and change your routine as the exam date gets closer. It will not help to change your routine and pull all-nighters right before the exam. In fact, if you pull an all-nighter the night before the exam, you may do much worse than you would have done if you kept your routine.

All-nighters or staying up late are not effective. For example, if you break your routine and stay up one-hour late, you will feel tired the next day. You may even have to sleep a few more hours the next day, adversely affecting your study regimen.

21. The importance of short, frequent breaks and physical exercise

Short, frequent breaks and physical exercise are VERY important for you, especially when you are spending a lot of time studying. They help relax your body and mind, making it much easier for you to concentrate when you study. They make you more efficient.

Take a five-minute break, such as a walk, at least once every one to two hours. Do at least 30 minutes of physical exercise every day.

If you feel tired and cannot concentrate, stop, go outside, and take a five-minute walk. You will feel much better when you come back.

You need your body and brain to work well to be effective with your studying. Take good care of them. You need them to be well-maintained and in excellent condition. You need to be able to count on them when you need them.

If you do not feel like studying, maybe you can start a little bit on your studies. Just casually read a few pages. Very soon, your body and mind will warm up and you will get into study mode.

Find a room where you will NOT be disturbed when you study. A good study environment is essential for concentration.

22. A strong vision and a clear goal

You need to have a strong vision and a clear goal: to master the architectural knowledge and become an architect in the shortest time. This is your number one priority. You need to master the architectural knowledge BEFORE you do sample questions or mock exams, except "truth-finding" exam or exercise at the very beginning of your exam prep. It will make the process much easier. Everything we discuss is to help you achieve this goal.

As I have mentioned on many occasions, and I say it one more time here because it is so important:

It is how much architectural knowledge and information you can <u>understand, digest, memorize</u>, and firmly retain that matters, not how many books you read or how many sample tests you have taken. The books and sample tests will NOT help you if you cannot understand, digest, memorize, and retain the important information for the ARE exam.

Cherish your limited <u>time and effort</u> and focus on the most <u>important</u> information.

23. Codes and standards used in this book
We use the following codes and standards:
American Institute of Architects, Contract Documents, Washington, DC; ADA Standards for Accessible Design, ADA; Various International Codes by ICC. See Appendixes for more information.

24. Where can I find study materials on architectural history?
Every ARE exam may have a few questions related to architectural history. The following are some helpful links to FREE study materials on the topic:
http://issuu.com/motimar/docs/history_synopsis?viewMode=magazine

Chapter Two

Practice Management (PcM) Division

A. General Information

1. **Exam content**

    The PcM division of the ARE has 80 questions which cover four different areas.

Sections	Target Percentage	Expected Number of Items
Section 1: Business Operations	20-26%	16-21
Section 2: Finances, Risk, & Development of Practice	29-35%	23-28
Section 3: Practice-Wide Delivery of Services	22-28%	17-23
Section 4: Practice Methodologies	17-23%	14-19

    Figure 2.1   Exam Content

    Note:
    After December 13, 2020, the number of questions will be reduced. See Figure 1.4.

    The exam content can be further broken down as follows:

    Section 1: Business Operations (20-26%)
    - Assess resources within the practice (A/E)
    - Apply the regulations and requirements governing the work environment (U/A)
    - Apply ethical standards to comply with accepted principles within a given situation (U/A)
    - Apply appropriate Standard of Care within a given situation (U/A)

    Section 2: Finances, Risk, & Development of Practice (29-35%)
    - Evaluate the financial well-being of the practice (A/E)
    - Identify practice policies and methodologies for risk, legal exposures, and resolutions (U/A)
    - Select and apply practice strategies for a given business situation and policy (U/A)

    Section 3: Practice-Wide Delivery of Services (22-28%)
    - Analyze and determine response for client services requests (A/E)
    - Analyze applicability of contract types and delivery methods (A/E)
    - Determine potential risk and/or reward of a project and its impact on the practice (A/E)

Section 4: Practice Methodologies (17-23%)
Analyze the impact of practice methodologies relative to structure and organization of the practice (A/E)
- Evaluate design, coordination, and documentation methodologies for the practice (A/E)
- Evaluate design, coordination, and documentation methodologies for the practice (A/E)

2. Official exam guide and reference index for the Practice Management (PcM) division

NCARB published the exam guides for all ARE 5.0 division together as *ARE 5.0 Handbook*.

You need to read the official exam guide for the PPD division at least three times and become very familiar with it. The real exam is VERY similar to the sample questions in the handbook.

You can download the official *ARE 5.0 Handbook* at the following link:
https://www.ncarb.org/sites/default/files/ARE5-Handbook.pdf

Note: We suggest you study the official ARE 5.0 Handbook first, and then other study materials, and then come back to Handbook again several days before the real ARE exam.

B. The Most Important Documents/Publications for Practice Management (PcM) Division of the ARE Exam

1. Official NCARB list of references for the Practice Management (PcM) division with our comments and suggestions

You can find the NCARB list of references for this division in the Appendixes of this book and the *ARE 5.0 Handbook*.

Note:
*While many of the MC questions in the real ARE exam **focus on design concepts**, there are **some questions requiring calculations**.*

In the ARE exams, it may be a good idea to skip any calculation question that requires over 2 minutes of your time; just pick a guess answer, mark it, and come back to calculate it at the end. This way, you have more time to read and answer other easier questions correctly.

A calculation question that takes 20 minutes to answer will gain the same number of points as a simple question that ONLY takes 2 minutes.

If you spend 20 minutes on a calculation question earlier, you risk losing the time to read and answer ten other easier questions, which could result in a loss of ten points instead of one.

The following is the NCARB list of top references for this division. For a longer list of relevant reference materials, please see the reference matrix at the end of this book.

Publications

The Architect's Handbook of Professional Practice
The American Institute of Architects
John Wiley & Sons, 14th edition (2008) and 15th edition (2014)

2020 Code of Ethics and Professional Conduct
AIA Office of General Counsel
The American Institute of Architects, 2020

Legislative Guidelines and Model Law/Model Regulations
National Council of Architectural Registration Boards
2018-2019

Model Rules of Conduct
National Council of Architectural Registration Boards
2018-2019

AIA Contract Documents
The following list of AIA Contract Documents have content covered in the Practice Management division.

B101-2017
Standard Form of Agreement Between Owner and Architect

C401-2017
Standard Form of Agreement Between Architect and Consultant

The following are some extra study materials that are useful if you have some additional time and want to learn more. If you are tight on time, you can simply look through them and focus on the sections that cover your weaknesses:

2. **Construction Specifications Institute (CSI) MasterFormat &** *Building Construction*
Become familiar with the new 6-digit CSI Construction Specifications Institute (CSI) MasterFormat as there may be a few questions based on this publication. Make sure you know which items/materials belong to which CSI MasterFormat specification section, and memorize the major section names and related numbers. For example, Division 9 is Finishes, and Division 5 is Metal, etc. Another one of my books, *Building Construction*, has detailed discussions on CSI MasterFormat specification sections.

Mnemonics for the 2004 CSI MasterFormat

The following is a good mnemonic, which relates to the 2004 CSI MasterFormat division names. Bold font signals the gaps in the numbering sequence.

This tool can save you lots of time: if you can remember the four sentences below, you can easily memorize the order of the 2004 CSI MasterFormat divisions. The number sequencing is a bit more difficult, but can be mastered if you remember the five bold words and numbers that are not sequential. Memorizing this material will not only help you in several divisions of the ARE, but also in real architectural practice

Mnemonics (pay attention to the underlined letters):
Good students can memorize material when teachers order.
F students earn F's simply 'cause **forgetting** principles have **an** effect. (21 and 25)
C students **end** everyday understanding things without memorizing. (31)
Please make professional pollution prevention inventions **everyday**. (40 and 48)

1-Good..................................General Requirements
2-Students..........................(Site) now Existing Conditions
3-Can..................................Concrete
4-Memorize.........................Masonry
5-Material ....Metals
6-When...............................Woods and Plastics
7-Teachers..........................Thermal and Moisture
8-Order...............................Openings

9-F......................................Finishes
10-Students.........................Specialties
11-Earn...............................Equipment
12-F's.................................Furnishings
13-Simply............................Special Construction
14-'Cause............................Conveying
21-Forgetting ....Fire
22-Principles........................Plumbing
23-Have..............................HVAC
25-An................................Automation
26-Effect.............................Electric

27-C...................................Communication
28-Students........................Safety & Security
31-End..............................Earthwork
32-Everyday.......................Exterior
33-Understanding ....Utilities
34-Things...........................Transportation
35-Without Memorizing........Waterways and Marine

40-Please..............................**Process Integration**
41-Make................................ Material Processing and Handling Equipment
42-Professional...................... Process Heating, Cooling, and Drying Equipment
43-Pollution.......................... Process Gas and Liquid Handling, Purification and Storage Equipment
44-Prevention........................Pollution Control Equipment
45-Inventions........................ Industry-Specific Manufacturing Equipment
48-Everyday..........................**Electrical Power Generation**

Note:
There are 49 CSI divisions. The "missing" divisions are those "reserved for future expansion" by CSI. They are intentionally omitted from the list.

Figure 2.2   Mnemonics for the 2004 CSI MasterFormat

Chapter Three

ARE Mock Exam for Practice Management (PcM) Division

A. **Multiple-Choice (MC)**

1. An owner hired a civil engineer to stake a building. The civil engineer could not find all the caisson dimensions to the property line on the structural plans, so he complained to the owner. The structural engineer is under the architect's contract. The owner asked the architect, and the architect informed the owner that the missing dimensions could be calculated by using the dimensions shown on the architectural and structural plans. The owner was not happy about this and wanted to hold the architect and structural engineer accountable. Which of the following is correct?
    a. Neither the structural engineer nor the architect is liable for this error.
    b. The structural engineer is liable for this error and should pay for it.
    c. Both the structural engineer and architect are liable for this error and should split the costs associated with the error.
    d. There is not enough information to decide.

2. An architectural firm has been awarded a contract to design a supermarket. The firm is in the process of building the project team. Choose the name of the employee on the following table who is the most appropriate project architect to be responsible for the overall design and development of the supermarket.

Name	Title	Licensed in Jurisdiction of Office Location	Licensed in Jurisdiction of Project Location	Years of Experience	Current Utilization Rate	Work Experience
Employee A	Principal/Partner	Yes	Yes	32	80%	Hospitality, Education, Retail
Employee B	Principal/Partner	Yes	No	26	50%	Residential, Education
Employee C	Architect	Yes	Yes	18	85%	Healthcare, Retail
Employee D	Architect	Yes	Yes	12	30%	Hospitality, Healthcare, Retail
Employee E	Architect	Yes	Yes	15	25%	Hospitality, Education
Employee F	Architect	Yes	No	8	70%	Hospitality, Education
Employee G	Architect	No	Yes	5	95%	Sports, Education
Employee H	Project Manager	Yes	Yes	15	50%	Hospitality, Retail
Employee I	Project Manager	No	Yes	10	90%	Healthcare
Employee J	Project Manager	No	No	7	93%	Hospitality, Education, Retail

Figure 3.1   Employee Workload

3. Which of the following insurances help to protect an architectural firm? Choose the two that apply.
    a. employment practice liability insurance
    b. business interruption insurance
    c. performance bond
    d. health insurance

4. See the following table. What is the utilization rate of Architect Frank?
    |____%____|

Architect Frank			
ID	Project Name	Task/Role	Average Weekly Hours
14836.01	Eastwood Hospital	Project Manager	16
16332.06	Grandview Apartment	Project Manager	12
18232.02	Parkwest Homes	Redlining	6
18663.00	Eastville Condominiums	RFP Team Coordination	2
17226.02	Office Management	Meeting with HR	2
18228.01	Continuing Education	Varies	2

Figure 3.2   Architect Frank Work Hours

5. An owner hired a surveyor to generate the boundary and topo survey plans. The architect designed a single-family home based on the survey plans. During the construction, the workers discovered an eighteen-inch public storm drain during the excavation. The city stopped the work and informed the owner and architect that the twenty feet wide easement for a storm drain was missing from the survey plans. Part of the building was placed inside the easement. As a result, the architect had to revise the plans that were approved by the city earlier. The design contract was based on AIA documents. Which of the following statements are correct? **Choose the two that apply.**
   a. The architect was liable for any damages caused by placing the building inside the easement.
   b. The architect was not liable for any damages caused by placing the building inside the easement.
   c. The architect should have revised the plans for free.
   d. The owner should have paid the architect extra to revise the plans.

6. If a client wants to know the project construction cost during the design stage, which of the following delivery methods should he choose? **Choose the two that apply.**
   a. design-build
   b. design-bid-build
   c. construction manager-constructor
   d. construction manager-agent

7. If an architect would like to protect his personal assets and maximize potential tax benefits, which legal entity should he establish for his practice? **Choose the two that apply.**
   a. sub C corporation
   b. sub S corporation
   c. limited liability partnership
   d. general partnership
   e. sole proprietorship

8. Which of the following are effective ways toward quality management in an architect's office? **Choose the two that apply.**
   a. establish a quality management program
   b. set quality requirements
   c. establish a project schedule
   d. use standardized details

9. An architectural firm needs to hire additional staff for a school project, what is the most important factor to consider for maximizing the profit of the project?
   a. the experience of the new hires in school design
   b. salary requirements of the new hires
   c. availability of the existing employees
   d. willingness of the existing employees to work overtime

10. An architect wants to set up a professional architectural corporation. Which of the following is *not* true?
    a. He needs to file an article of incorporation with the applicable secretary of state.
    b. All the directors, officers, and stockholders must be licensed architects in the state in which they practice.
    c. He needs to file an extra copy of the article of incorporation with the internal revenue service.
    d. He needs to obtain a business license from the city.
    e. The professional architectural corporation can be a subchapter S corporation.
    f. The professional architectural corporation owners remain personally liable for personal negligence and wrongdoing.

11. What is an effective way of prohibiting moonlighting by employees? **Choose the two that apply.**
    a. noncompete clause
    b. inclusion of the requirement in the employee manual
    c. verbal agreement
    d. employment contract
    e. at-will employment clause

12. An architectural firm with sixty employees is looking for a project architect to work on a project in China. Which of the following questions can the interviewers ask? **Choose the two that apply.**
    a. What experience do you have in your previous job?
    b. Are you from China?
    c. Do you speak Chinese?
    d. Do you have any physical problems that would prevent you from doing the job?

13. An architectural firm is responding to a Request for Proposal (RFP) for a school project. The firm's principal had previously worked on six school projects with his former employer, a competing firm. The principals are both AIA members. What is the best way to include these six projects in the RFP?

a. Contact the previous employer to obtain permission to use the photos for these six projects.
b. Use the project photos, and in the project description, give credit to the other firm.
c. List the project as experience, but clearly indicate the project was done while working for the other firm.
d. Do not use the projects at all to avoid potential problems.

14. A student sent a resume to an architectural firm looking for an internship opportunity and has indicated she is willing to work for free. What should the architect's response be?
    a. Hire the student on an unpaid basis.
    b. Tell the student the firm has to pay her if she is hired.
    c. Hire the student, pay her, and advise her to start NCARB AXP.
    d. Tell the student that if the firm decides to retain her for an internship, she will be paid, and advise her to start NCARB AXP.

15. A developer asked an architect to design a supermarket. After the architect finished the design development plans, the project was stopped for lack of funding. Three years later, the developer came back and asked the architect to finish the design. The building codes were modified one year after the project was stopped. What should the architect do?
    a. Finish the project design per the original program and ignore the building code updates.
    b. Finish the project design per the original program and incorporate the new building code requirements without extra charges to the client.
    c. Finish the project design per the original program and incorporate the new building code requirements with extra charges to the client.
    d. Send the client a proposal requesting additional design fees to cover incorporating the new building code requirements into the set and proceed upon approval of the client.

16. A famous home and garden magazine requests to publish an article about an architect's completed project. The magazine wants to publish the detailed floor plans and construction cost in addition to photographs. Per *Code of Ethics and Professional Conduct* by AIA Office of General Counsel, how can the architect respond to this request?
    a. The architect can provide the information unless the owner requests the architect not to do so.
    b. The architect needs to obtain written permission from the owner before providing the information.
    c. The architect can provide the detailed floor plans, but not the construction cost.
    d. The architect can provide the detailed floor plans and the photographs only.

17. Which of the following has the most influence on the financial well-being of a firm?
    a. payroll costs
    b. aged account receivables
    c. federal and state taxes
    d. insurance costs

18. Which of the following can help an architect to minimize the claims by subcontractors?
    a. Follow proper protocol and make sure that all communications with the subcontractors go through the general contractor.
    b. Communicate with the subcontractors directly to avoid confusion.
    c. Obtain the owner and the general contractor's approval before communicating with the subcontractors directly to avoid confusion.
    d. Include an indemnification clause in the agreement.

19. What is the best way to protect an architect's copyright for his work?
    a. Sign a separate copyright agreement with the client.
    b. Use standard AIA documents for agreements.
    c. Add a copyright notice to all plans and project manuals. Also register all works with the US copyright office.
    d. Do nothing since all architect's works are creative works and are protected automatically by copyright laws.

20. A balance sheet is important for evaluating a firm's financial well-being because it
    a. shows the percentage of billable hours vs. non-billable hours
    b. shows an overall picture of the firm's assets and liabilities
    c. shows the profit and loss of the firm
    d. shows the percentage of billable hours for management and technical staff

21. What range should the current ratio be?
    a. 1 to 1.5
    b. 1.5 to 2
    c. 2.5 to 3
    d. 3 to 3.5

22. An architectural firm has many repeat clients. What is the best way to maintain a good relationship with the clients while still making a profit and maximizing high-quality service? **Choose the two that apply.**
    a. Tell the clients that the architect will provide the highest quality of service and charge them the lowest fees.
    b. Tell the clients that the architect will provide the average quality of service and charge them reasonable fees.
    c. Increase the firm's service fees each year based on CPI.
    d. Increase the firm's service fees every three years based on CPI and building code updates.
    e. Try to keep the firm's service fees low, and only increase when necessary.

23. What is a net multiplier? **Choose the two that apply.**
    a. the ratio of net revenue to direct salary expenses
    b. the ratio of net revenue to total expenses
    c. a common way to evaluate staff productivity
    d. a common way to evaluate overhead staff productivity

24. Which of the following is a tort?
    a. intentional concealment of material facts by an architect
    b. unauthorized downloading and use of a CADD software in an architectural firm
    c. negligence by the architect
    d. theft of tools from the project site

25. What is a utilization ratio?
    a. the sum of usable areas divided by the total area of a building
    b. the sum of programmed spaces divided by the total spaces
    c. a measure of the efficiency of a mechanical system
    d. the sum of building areas divided by the site area
    e. the sum of billable hours divided by the total hours of an employee

26. What is a quick ratio?
    a. the actual salary of an employee divided by the total cost of the employee including salary, fringe benefits, and taxes, etc.
    b. the total current assets and unbilled revenue divided by the total of current liabilities
    c. the costs of personnel divided by the sum of the costs of personnel and overhead
    d. the costs of labor divided by the sum of the costs of labor, overhead, and profit

27. Which of the following regarding the joinder provisions in AIA documents are true? **Choose the two that apply.**
    a. They allow parties to consolidate arbitrations.
    b. They allow parties to include a third party by joinder in arbitrations.
    c. They allow parties to join as a team to split legal costs.
    d. They allow an architect to be the initial decision maker in arbitration.

28. Which of the following are true regarding a project alliance? **Choose the two that apply.**
    a. Parties waive liabilities against each other.
    b. This practice is often used in integrated project delivery (IPD).
    c. A project alliance is a single-purpose legal entity, similar to a corporation.
    d. A project alliance is a form of design-build.
    e. Project teams make a consensus-based agreement.

29. An owner is a close friend of an architect. She asks the architect for advice concerning owner's insurance for a project. What are the proper responses by the architect? **Choose the two that apply.**
    a. Refer her to her own insurance agent or attorney for advice.
    b. Recommend the general liabilities insurance, worker's comp insurance, performance bond, and other typical insurances required for a project and ask her to verify the information herself.
    c. Give her the AIA document G612, Owner's Instructions to the Architect Regarding the Construction Contract.
    d. Set up a meeting between the owner and the architect's insurance agent.

30. A very busy small architectural firm has obtained a major project. What are the likely changes the firm will make? **Choose the two that apply.**
    a. Hire new staff as needed.
    b. Hire new staff quickly to handle the new project.
    c. Ask the existing staff to work overtime.
    d. Work with the client to adjust the deadline.

31. What are the best ways to minimize risks for an architectural firm? **Choose the two that apply.**
    a. Start the effort of minimizing risks at the feasibility study stage.
    b. Start the effort of minimizing risks at the marketing stage.
    c. Do research and interview developers.
    d. Select clients with a similar value.
    e. Prepare the contract between the architect and client.

32. Which of the following business organizations will minimize the risk for the owner and principals of a firm?
    a. sole proprietorship
    b. limited partnership
    c. joint venture
    d. limited liability company
    e. professional corporation

33. A studio-based architectural firm with twenty employees is expanding into new service areas of interior design and LEED certification. Which of the following is the best organization plan for the expansion?
    a. Keep the current studio structure.
    b. Create specialty departments.
    c. Utilize studios with experience in the project types.
    d. Convert to a horizontal departmental structure.

34. What is the best way to get the client involved in the early stages of design?
    a. phone calls
    b. questionnaires sent to the client
    c. PowerPoint presentations
    d. hand sketches
    e. a design charrette

35. The architect has regularly exchanged CADD files with her consultants. The electrical contractor has installed the lights per the electrical plans, but the light locations are different from the ones shown on the architectural plans and are in conflict with the HVAC plans. Per C401-2007, Standard Form of Agreement Between Architect and Consultant, who should pay for the corrections?
    a. architect
    b. electrical engineer
    c. mechanical engineer
    d. general contractor
    e. electrical contractor

36. What is the most cost-effective way for an eighteen-person firm specializing in retail buildings to create project specifications? **Choose the two that apply.**
    a. Use an in-house specifications writer to modify the office's prototype specifications.
    b. Subscribe to master specifications and modify them per the specific project.
    c. Hire an outside specifications consultant.
    d. Ask the project manager for each project type to develop specifications per the office's prototype specifications.

37. While designing a six-story court house and detention facility complex, which of the following individuals should an architect most actively involve regarding vertical transportation?
    a. elevator consultant
    b. structural engineer
    c. electrical engineer
    d. mechanical engineer
    e. plumbing engineer
    f. security consultant
    g. facility engineer
    h. owner

38. A client with no construction experience wants to build his dream home. He comes to an architect with a budget of at least 30% below the actual construction cost, and 20% below the needed architectural fees. What should the architect do?
    a. Decline the offer and tell him that the budget is not feasible for what he wants to build.
    b. Decline the offer and tell him it is a violation of the AIA *Code of Ethics and Professional Conduct* to accept a project that is clearly under budget.
    c. Accept the offer but ask the client to increase the budget.
    d. Accept the offer and strive to meet the budget in an innovative way.

39. Which of the following should an architect know about a client in order to minimize project risks? **Choose the four that apply.**
    a. personality and character
    b. prior project history
    c. financial ability
    d. their ideas about how the building should look
    e. the client's project team
    f. the client's preferred type of service agreement

40. Design-assist contracting is
    a. a form of architect as owner's adviser project delivery
    b. a form of architect as contractor's adviser project delivery
    c. a form of project delivery which involves specialty subcontractors or trades early in the design and construction document phase
    d. a form of architect as owner's construction manager project delivery

41. An architect designs a building for an owner. Which of the following are true? **Choose the two that apply.**
    a. The architect owns the copyright for the plans of the project.
    b. The owner owns the copyright for the plans of the project.
    c. The architect owns the copyright for the actual building.
    d. The owner owns the copyright for the actual building.
    e. There is a copyright for the plans, but no copyright for the actual building.

42. An electrical contractor discovered a major error on the electrical drawings during construction. The owner asked the architect to pay for the correction. The architect in turn asked the electrical engineer to pay for it. The electrical engineer refused to pay for it. What should the architect do? **Choose the two that apply.**
    a. Inform the owner that the electrical engineer should pay for it.
    b. Send a formal payment request to the electrical engineer and copy the owner in the correspondence.
    c. Pay for the correction.
    d. Sue the electrical engineer for the cost of the correction plus the legal costs.

43. Which of the following are true? **Choose the two that apply.**
    a. An architect should always use AIA documents.
    b. An architect can use ConsensusDOCS.
    c. An architect can use AIA documents or ConsensusDOCS.
    d. An architect should use ConsensusDOCS.

44. If an architect uses an outside drafting consultant, she needs to keep documents of the drafter's involvement and architect's supervision of the work for:
    a. four years
    b. five years
    c. seven years
    d. ten years
    e. four to fifteen years, depending on the state she lives in

45. What is the best way to modify an AIA document?
    a. Retype the document and then revise.
    b. Cross out the text by hand and write in the changes by hand.
    c. Have an attorney draft a new contract using some of the language from the actual AIA document.
    d. Attach amendments or supplemental conditions.

46. Which of the following fee structures will encourage a contractor to be most efficient?
    a. cost-plus with a fixed fee for overhead and profit
    b. cost-plus with overhead and profit based on a percentage of the construction cost
    c. unit price
    d. stipulated sum

47. An architect is liable for any errors and omissions of her consultants. This kind of indirect liability is called _____.
    a. vicarious liability
    b. professional liability
    c. limited liability
    d. transferred liability

48. AIA Document A201-2007, General Conditions of the Contract for Construction requires the owner to buy and keep builder's all-risk insurance or equal. Which of the following are excluded from an all-risk insurance? **Choose the two that apply.**
    a. act of war
    b. employee theft
    c. collapse
    d. theft
    e. debris removal
    f. vandalism
    g. explosions

49. A client submitted a set of standard blueprints purchased from a homebuilding publisher to the city for a building permit. The application was denied, and the city requested more information such as structural plans, and energy calculations, etc. The client comes to an architect and asks her for assistance to obtain the permit. What should the architect do? **Choose the two that apply.**
    a. Give the client a quote to provide only the missing information needed to obtain the permit.
    b. Politely decline the request since it is not cost effective to do the project.
    c. Politely decline the request since it is against the NCARB's *Model Rules of Conduct* and copyright laws to modify another designer's work.
    d. Send the client a proposal to develop a new set of plans using the standard design as a starting point.

50. An architect should keep project records
    a. for four years after substantial construction
    b. until the end of the statute of limitations
    c. until the end of the statute of repose
    d. for four years after substantial completion

51. An owner uses design development plans to obtain a fixed price bid, and then allows the contractor to decide the details of the construction. This method of construction delivery is called
    a. bridging
    b. design-build
    c. design-bid-build
    d. integrated project delivery
    e. fast-track

52. Which of the following are true regarding the construction management at risk (CMAR) project delivery method? **Choose the two that apply.**
    a. The construction documents must be completed before construction starts.
    b. It has two separate contracts.
    c. It has three prime players.
    d. It has three prime contractors.

53. A supermarket owner wants to make sure the market can have a grand opening within one year. What is the best project delivery method?
    a. design-build
    b. design-bid-build
    c. fast track
    d. negotiated bid with a guaranteed maximum price (GMP)

54. A client asks the architect to reduce the time needed for design and construction documents after the contract is signed, what is the best course of action for the architect? **Choose the two that apply.**
    a. Ask the client to pay additional fees.
    b. Hire new employees to speed up the project.
    c. Assign more existing staff to the project.
    d. Ask the client to reduce the scope of the project.
    e. Ask the project team to work overtime.

55. An owner failed to pay the architect per the contract. The architect followed the proper procedure and terminated the contract. The owner tried to continue to use the construction documents to finish the project. What is the proper action for the architect?
    a. Do nothing.
    b. Send a written notice asking the owner to pay for the services rendered and informing him that the architect will take legal action if not reimbursed promptly.
    c. Send a written notice asking the owner to cease and desist.
    d. none of the above

56. According to the MacLeamy curve, an architect should allocate the most resources during which phase of the project using integrated project delivery (IPD)?
    a. conceptualization
    b. criteria design
    c. detailed design
    d. implementation document
    e. agency coordination
    f. construction

57. Two architectural firms are considering teaming up to bid on a large project. What kind of agreement should they use?
    a. joint venture agreement
    b. prime consulting agreement
    c. teaming agreement
    d. corporation agreement

58. Which of the following are mandatory requirements for NCARB's Architectural Experience Program (AXP)? **Choose the four that apply.**
    a. a licensed architect as supervisor
    b. a NCARB record
    c. experience in specific categories
    d. online reporting
    e. state coordinators
    f. interactive online service

59. An architectural firm wants to change its at-will employment policy to employment contract. What is the biggest impact on the firm?
   a. increased book keeping requirements
   b. the need to give a reason for terminating an employee
   c. moonlighting is prohibited
   d. creating specific job descriptions for each position

60. Which of the following accounting methods should a twelve-person architectural firm use?
   a. accrual basis
   b. cash basis
   c. double entry
   d. modified accrual basis

61. An architect is preparing a proposal for an inexperienced client. Which of the following is the best pricing method?
   a. cost-plus fee
   b. percentage of construction cost
   c. hourly not to exceed
   d. square-feet cost

62. If an architectural office's overhead rate is 1.7, what should the architect do to reduce the overhead? **Choose the three that apply.**
   a. Review reimbursables.
   b. Monitor unbillable hours more closely.
   c. Be vigilant in fee collection.
   d. Increase revenue per technical staff.
   e. Move to a cheaper office.
   f. Ask the administrative staff to check their direct expenses.

63. Which of the following reports should an architect use to evaluate the financial performance of her firm?
   a. profit and loss statement
   b. net profit before tax
   c. balance sheet
   d. office earning report

64. An architectural firm specializes in office buildings is trying to expand its services to retail buildings. Which of the following are proper strategies? **Choose the two that apply.**
   a. Develop a web page on the firm's website showcasing retail building design.
   b. Start social media marketing.
   c. Research the market and identify possible contacts.
   d. Hire a professional with experience in retail buildings and office buildings.

65. Which of the following are most effective for quality control during construction document production? **Choose the four that apply.**
    a. Assign one person to be in charge of the project.
    b. Hire an outside firm to do final quality control.
    c. Exchange progress plans with the consultants regularly.
    d. Use a construction specific institute (CSI) checklist.
    e. Use office standard details.
    f. Use CSI's Uniform Drawing System (UDS) of National CAD Standard (NCS).

B. Case Study

Case Study 1

Questions 66 through 75 refer to the following case study. See below for information necessary to answer the questions.

An architectural firm is a full-service practice with two principals, three project managers, five project architects, one specification writer/material specialist, one contract/construction administrator, one office manager/accountant, one marketing manager, one receptionist/administrative assistant, and eight drafters. The firm's technical staff is familiar with standard computer-aided design (CAD) software. The firm's principals are seriously considering converting to Building Information Modeling (BIM). However, most of the staff is not familiar with BIM.

The firm has secured commissions for a 30,000-sf supermarket, and fifteen fast-food chain restaurants. Both projects use standard AIA documents. Both clients provided their prototype plans.

66. The firm carries a one-million-dollar professional liability insurance, and the owner of the supermarket wants the architect to increase the coverage of the policy to three million dollars. Who should pay for the extra premium?
    a. architect
    b. owner
    c. owner or architect, depending on the agreement
    d. owner and architect should split the extra cost equally

67. The existing staff of the firm is busy with many other projects, and they do not have enough time to handle the two new projects. What is the best solution for the firm's principals?
    a. Find another firm to form a joint venture.
    b. Hire an outside freelance consultant.
    c. Sign a prime contract agreement with another architectural firm.
    d. Sign a teaming agreement with a local firm.

68. The architect recommends a construction contract with a guaranteed maximum price to the owner of the fast-food restaurant chain. The owner asks the architect to guarantee that the project will not be over this budget. The architect agrees to the owner's request. In doing so, the architect has?
    a. violated the anti-trust laws
    b. voided his error and omission professional liabilities insurance
    c. violated the AIA *Code of Ethics and Professional Conduct*
    d. raised the level of standard care

69. What types of consultants should be retained to complete the fast-food restaurant projects? **Choose the four that apply.**
    a. fixture
    b. structural
    c. mechanical
    d. electrical
    e. plumbing
    f. signage
    g. interior designer
    h. egress
    i. security

70. During the bidding stage of the supermarket project, one bidder discovers some major discrepancies between the mechanical and electrical plans and informs the architect. The bid is due in one week. What should the architect do?
    a. Finish the bidding process and ask the engineers to coordinate and correct the plans after the bidding.
    b. Ask the engineers to coordinate and correct the plans and issue the corrected plans to the bidder who brought up the issue.
    c. Ask the engineers to coordinate and correct the plans and issue the corrected plans to all bidders.
    d. Inform the owner of the situation and ask if the bid deadline can be postponed so that the revised plans can be issued to all bidders with additional time for the bidders to modify their bids accordingly.

71. During a field visit to the construction site of the supermarket, the architect notices that the plywood sheathing is loosely placed on the roof, not nailed down, and has no weather protection. The architect brings this issue up to the contractor and owner in writing. However, the contractor thinks rain is unlikely and does nothing. A few days later, a heavy storm comes through, the plywood sheathing gets wet, and many of the pieces become warped and unusable. Who should bear the cost of replacing the damaged plywood sheathing?
    a. owner
    b. architect
    c. contractor
    d. framing subcontractor

72. During construction, the mechanical contractor discovers that there is not enough clearance space between the mezzanine floor joists and the first-floor ceiling to install the mechanical ducts as planned. He has to return the specified mechanical ducts and order different-sized ducts after the architect does the applicable research and issues revised plans. The mechanical contractor sends a bill to the architect and asks the architect to pay for the extra costs caused by this revision. What legal concept can protect the architect from the claim by the mechanical contractor?
    a. agency
    b. fiduciary duty
    c. lack of privity
    d. indemnification

73. Because of the problems with these two projects and previous projects, the architectural firm is trying to improve its methods of document production. The firm already has competent staff and a standard detail library. What other method may be beneficial to the firm?
    a. Ask one of the principals to do quality control of the documents.
    b. Ask the project manager to be more careful and spend more time checking the documents.
    c. Hire an outside firm to check the documents.
    d. Develop a project check list.

74. The corporation that owns fifteen fast-food chain restaurants has a set of prototype plans, and has required the architect to design all the buildings with a look as close to the prototype as possible. The cities where these projects are located have different planning guidelines, and some even require a certain architectural style depending on location. How should the architect proceed?
    a. Design the buildings according to the client's prototypes.
    b. Design the buildings according to the cities' planning guidelines.
    c. Design the buildings according to both the client's prototypes and the cities' planning guidelines.
    d. Design the buildings according to the client's prototypes, submit them for the cities' review, and apply for variances as needed.

75. Besides retail projects, the architectural firm also wants to focus on office buildings. What is the best way to organize the office?
    a. departmental
    b. departmental with two design specialists
    c. a single studio
    d. two specialist studios

Chapter Three

Case Study 2

Questions 76 through 80 refer to the following case study. See below for information necessary to answer the questions.

An architectural firm is working on a 60,000-sf shopping center with a big box retail building and a two-story shop building. The project is in the construction phase. A fire burnt down the two-story shop building, and only the slab, some steel beams, and columns are left. The owner wants the architect to secure any necessary approvals to rebuild the burnt down building.

76. Who should pay for the fire damages of the burnt down building?
    a. owner
    b. contractor
    c. architect
    d. owner's insurance company
    e. contractor's insurance company

77. The structural engineer needs to send staff to check if the remaining steel beams and columns after the fire are still usable. Who should pay for the extra field visits?
    a. owner
    b. architect
    c. contractor
    d. structural engineer

78. During construction of the big box retail building, an electrician fell to his death while trying to install a ceiling light. Who should pay for this accident?
    a. owner
    b. contractor
    c. architect
    d. owner's insurance company
    e. contractor's insurance company

79. At the end of construction, the owner requests to occupy part of the big box retail building before it is completed. Whose approval should the owner get before he can move in? **Choose the two that apply.**
    a. city
    b. owner
    c. contractor
    d. architect
    e. owner's insurance company
    f. contractor's insurance company

80. At the end of construction, the owner asks the architect to explore the possibilities of getting LEED certification for the project. Which of the following LEED certifications is most likely to be achieved? **Choose the two that apply.**
   a. LEED BD+C
   b. LEED ID+C
   c. LEED O+M
   d. LEED Retail

Chapter Four

ARE Mock Exam Solutions for Practice Management (PcM) Division

A. Mock Exam Answers and Explanations: Multiple-Choice (MC)

Note: If you answer 68% of the questions correctly, you pass the MC Section of the exam.

1. Answer: a
   Neither the structural engineer nor the architect is liable for this error.

   According to the *Architect's Handbook of Professional Practice*, it is not realistic to expect the architect and his consultants to be perfect and not make any errors. The architect just needs to perform reasonably and prudently in order to meet the professional standard of care for architects.

   Since the missing dimensions can be calculated by using the dimensions shown separately on the architectural and structural plans, strictly speaking, this is not even an error.

2. Answer: Employee D
   Employees D and E both have very low current utilization rate and are licensed in the jurisdiction where the project is located. However, Employee E has no experience in retail projects. Therefore, Employee D is the most appropriate project architect to be responsible for the overall design and development of the supermarket.

3. Answer: a and b
   According to the *Architect's Handbook of Professional Practice*, the following insurance types help protect an architectural firm:
   - **Employment practice liability insurance** protects an architectural firm when employees sue the firm for charges like harassment, discrimination, and wrongful termination, etc.
   - **Business interruption insurance** reimburses the firm for loss of profits and continuing fixed expenses for business interruption caused by fire, windstorm, computer crashes, etc.

   The following are incorrect answers:
   - **A performance bond** protects the owner during construction in case the contractor does not finish the work.
   - **Health insurance** protects employees, not the firm specifically.

4. Answer: The utilization rate of Architect Frank is
   | 85 % |

| Architect Frank ||||
ID	Project Name	Task/Role	Average Weekly Hours
14836.01	Eastwood Hospital	Project Manager	16
16332.06	Grandview Apartment	Project Manager	12
18232.02	Parkwest Homes	Redlining	6
18663.00	Eastville Condominiums	RFP Team Coordination	2
17226.02	Office Management	Meeting with HR	2
18228.01	Continuing Education	Varies	2

Figure 4.1   Architect Frank Work Hours

According to the *Architect's Handbook of Professional Practice* and *Architect's Guide to Small Firm Management*, utilization rate is calculated by dividing an employee's direct labor (billable hours) by their total labor (total hours).

Time spent on Eastwood Hospital, Grandview Apartment, and Parkwest Homes is all direct labor because it is billable to a specific project (**Step 1**).

Hours spent on RFP team coordination, meeting with HR, and continuing education are considered indirect labor (**Step 2**).

Dividing the direct labor hours by the total hours will find Architect Frank's utilization rate of 85%. (**Step 23**).

Step 1: 16 + 12 + 6 = 34 hours
Step 2: 16 + 12 + 6 + 2 + 2 +2 = 40 hours
Step 3: 34/40 = 0.85 or 85%

5. Answer: b and d
   Since the owner hired a surveyor directly to generate the boundary and topo survey plans, per B101-2007, Standard Form of Agreement Between Owner and Architect, the architect should be entitled to rely on the accuracy and completeness of the services and information furnished by the owner and the owner's consultants. Therefore, the architect was not liable for any damages caused by placing the building inside the easement, and the owner should have paid the architect extra to revise the plans.

6. Answer: a and c
   Per *Professional Practice: A Guide to Turning Designs into Buildings* and *The Architect's Handbook of Professional Practice*, if a client wants to know the project construction cost during the design stage, he should choose the following delivery methods:
   - **Design-build** has a single contract between the owner and the design-build entity. The contract includes a fixed price for both design and construction cost.
   - In the **construction manager-constructor** delivery method, the contractor can get involved early in the design stage, and provide a guaranteed maximum price based on early design documents.

   The client should not choose the following delivery methods:
   - **Design-bid-build** is the conventional approach. The owner will not know the project's construction cost until after the contract documents are completed and bid.
   - **Construction manager-agent** can provide the owner with early construction input, but this delivery method does not provide a final construction cost in the design phase.

7. Answer: b and c
   If an architect would like to protect his personal assets and maximize potential tax benefits, he should establish one of the two following legal entities for his practice:
   - Per *Professional Practice: A Guide to Turning Designs into Buildings* and *The Architect's Handbook of Professional Practice*, partners of a **limited liability partnership (LLP)** may be liable for their own personal negligence but are not personally responsible for liabilities incurred by the LLP. LLP's are not required to pay federal taxes, because income and losses are reported on each individual's tax return.
   - **Sub S corporation** practices protect stockholders' personal assets. The corporation income can be passed through to stockholders and be claimed on personal tax returns to avoid double taxation.

   He should *not* establish the following legal entities for his practice:
   - **Sub C corporations** have the problem of double taxation. Sub C corporations protect stockholders' personal assets. However, both the corporation and stockholders pay federal taxes.
   - **General partnership** provides no liability protection for personal assets. Partners have total liability and their personal assets are liable to the partnership's obligations.
   - **Sole proprietorship** provides no liability protection for personal assets.

8. Answer: a and b
   Per *the Architect's Handbook of Professional Practice*, the following are effective ways toward quality management in an architect's office:
   - establish a quality management program
   - set quality requirements

The following are not the best answers:
- Establishing a project schedule helps the management of the project, but has no direct impact on quality management.
- Using standardized details saves time, but it requires modifications of the details. If the project team fails to modify the details per the project's specific requirements, it will have a negative impact on quality management.

9. Answer: a

The most important factor to consider for maximizing the profit of the project is the experience of the new hires in school design.

The following are not the best answers
- salary requirements of the new hires; A lower-paid employee may not make the project more profitable because he may take longer to complete the task, and other staff may have to correct his mistakes. Many architectural firms learn this the hard way.
- availability of the existing employees; Obviously the existing employees are very busy, otherwise the firm would not be hiring new staff.
- willingness of the existing employees to work overtime; Asking the existing employees to work overtime is not the best way to save money because overtime pay is 150% of regular pay. Also, employees working overtime are tired and cannot work efficiently.

10. Answer: c

Please pay attention to the word "not."

If an architect wants to set up a professional architectural corporation, the following is *not* true:
- He needs to file an extra copy of the article of incorporation with the internal revenue service.

All the other answers are correct.
- He needs to file an article of incorporation with the applicable secretary of the state.
- All the directors, officers, and stockholders must be licensed architects in the state in which they practice. This requirement is different for a regular corporation. A professional architectural corporation can provide professional services while a regular corporation cannot.
- He needs to obtain a business license from the city.
- The professional architectural corporation can be a subchapter S corporation. An architect operating under a professional corporation or a C corporation can also select treatment by the IRS as a personal services corporation, which the IRS defines as a corporation with the primary activity of providing personal services performed mainly by the corporation's employee owners.
- The professional architectural corporation owners remain personally liable for personal negligence and wrongdoing.

11. Answer: b and d
    The following are effective ways of prohibiting moonlighting by employees:
    - inclusion of the requirement in the employee manual; The employer can ask each new hire to sign a statement that he has received the employee manual and will obey its requirements as a condition of employment.
    - employment contract; If clearly written, this rule can be easily enforceable.

    The following are *not* effective ways of prohibiting moonlighting by employees:
    - A noncompete clause will prohibit employees from competing with the firm after their employment, not during employment.
    - Verbal agreements are legally binding but very hard to enforce.
    - An at-will employment clause simply means that either the employer or the employee can terminate the employment at any time, with or without any reason. It cannot prevent the employee from moonlighting.

12. Answer: a and c
    The interviewers can ask the following questions:
    - What experience do you have in your previous job? (This is related to the job.)
    - Do you speak Chinese? (This is related to the job since the project is in China.)

    The interviewers *cannot* ask the following questions:
    - Are you from China? (Equal employment opportunity laws prohibit employers from asking about national origin, marital status, race, and age.)
    - Do you have any physical problems that would prevent you from doing the job? (ADA does not allow firms with fifteen or more employees to ask questions about physical problems, including those that may interfere with doing the job.)

13. Answer: c
    Per *Code of Ethics and Professional Conduct* by AIA Office of General Counsel, the best way to include these six projects in the RFP is to:
    - List the project as experience, but clearly indicate the project was done while working for the other firm.

    The follow choices are not the best:
    - Contact the previous employer to obtain permission to use the photos for these six projects. (Using the photos can be embarrassing, especially if the other firm also uses the same project photos.)
    - Use the project photos, and in the project description, give credit to the other firm. (Using the photos can be embarrassing, especially if the other firm also uses the same project photos.)
    - Do not use the projects at all to avoid potential problems. (AIA *Code of Ethics and Professional Conduct* does allow an architect to honestly cite his experience with other firms and claim credit properly.)

14. Answer: d
    Per *Code of Ethics and Professional Conduct* by AIA Office of General Counsel,
    The architect's response should be:
    - Tell the student that if the firm decides to retain her for an internship, she will be paid, and advise her to start NCARB AXP.

    The following are incorrect answers:
    - Hire the student on an unpaid basis. (This does not comply with *Code of Ethics and Professional Conduct.*)
    - Tell the student the firm has to pay her if she is hired. (The architect should also advise her to start NCARB AXP.)
    - Hire the student, pay her, and advise her to start NCARB AXP. (The firm still needs to interview and evaluate the application package of the student and may decide not to hire her.)

15. Answer: d
    Per B101, Standard Form of Agreement Between Owner and Architect, the architect should:
    - Send the client a proposal requesting additional design fees to cover incorporating the new building code requirements into the set and proceed upon approval of the client.

    The following are incorrect answers:
    - Finish the project design per the original program and ignore the building code updates. (This is not acceptable standard of care for an architect.)
    - Finish the project design per the original program and incorporate the new building code requirements without extra charges to the client. (Building codes updates are considered additional services that require additional fees.)
    - Finish the project design per the original program and incorporate the new building code requirements with extra charges to the client. (Per B101, Standard Form of Agreement Between Owner and Architect, the architect should *not* proceed until she receives the owner's written authorization.)

16. Answer: a
    Per *Code of Ethics and Professional Conduct* by AIA Office of General Counsel, the architect should maintain the confidentiality of the owner when requested. After the architect notifies the owner of the magazine's request, the owner has the opportunity to ask the architect not to publish the information. The architect can provide the information unless the owner prompts the architect not to do so.

    Per AIA document B141, *Standard Form of Agreement between Owner and Architect with Standard Form of Architect's Services,* an architect has the rights to take photographs of the project and use them in promotion and professional materials, unless the owner has previously informed the architect, in writing, of what specific information is confidential.

17. Answer: b
Aged account receivables have the most influence on the financial well-being of a firm. Getting paid on time is critical and will insure proper cash flow of the firm. A firm's management team needs to know the status of aged account receivables. Invoices that have not been paid in a timely manner (typically within sixty days) should be brought to the attention of the principal.

- Payroll costs refer to salary or wages and the related federal and state employment taxes.
- Federal and state taxes are standard items that all firms need to deal with.
- Insurance costs are important but not as critical as aged account receivables.

18. Answer: d
Including an indemnification clause in the agreement can help an architect to minimize the claims by subcontractors.

The following are incorrect answers:
- Follow proper protocol and make sure that all communications with the subcontractors go through the general contractor. (This will help to reduce confusion, but will not minimize the claims by subcontractors.)
- Communicate with the subcontractors directly to avoid confusion. (An architect should never communicate with the subcontractors directly; all communications with the subcontractors should go through the general contractor.)
- Obtain the owner and the general contractor's approval before communicating with the subcontractors directly to avoid confusion. (See above.)

19. Answer: c
The best way to protect an architect's copyright for his work is to add a copyright notice to all plans and project manuals as well as register all works with the US copyright office.

The following are incorrect answers:
- Sign a separate copyright agreement with the client (This is not necessary since AIA documents already include copyright provisions.)
- Use standard AIA documents for agreements (This is a good idea, but the architect should also add a copyright notice to all plans and project manuals. In addition, registering all works with the US copyright office will enable an architect to seek attorney's fees if a lawsuit is filed to protect the copyright and seek damages.)
- Do nothing since all architect's works are creative works and are protected automatically by copyright laws. (This statement is true. However, this is not the best way to protect an architect's copyright for his work. Without registering all works with the US copyright office, it is hard to enforce your copyright.)

The Architectural Works Copyright Protection Act allows an architect to copyright an *actual* building to prevent others from copying his building design.

20. Answer: b
A balance sheet is important for evaluating a firm's financial well-being because it shows an overall picture of the firm's assets and liabilities.

The following are incorrect answers:
- shows the percentage of billable hours vs. non-billable hours (This information is found on an individual's timesheet.)
- shows the profit and loss of the firm (This information is located on the profit and loss statement.)
- shows the percentage of billable hours for management and technical staff. (This information is found in the summary of the firm's timesheets.)

21. Answer: a
The range of the current ratio should be 1 to 1.5. In general, higher numbers represent better financial well-being for the firm.

The **current ratio** is a liquidity ratio that measures if a firm has enough resources to meet its short-term obligations. It is expressed as follows:

$$Current\ Ratio = \frac{Current\ Assets\ (both\ liquid\ and\ non-liquid)}{Current\ Liabilities}$$

The current ratio indicates a firm's liquidity.

22. Answer: b and c
The best way to maintain a good relationship with the clients while still making a profit and maximizing high-quality service are to:
- Tell the clients that the architect will provide the average quality of service and charge them reasonable fees.
- Increase the firm's service fees each year based on CPI.

The following are incorrect answers:
- Tell the clients that the architect will provide the highest quality of service and charge them the lowest fees. (Never do this because it will generate unnecessary liabilities for the firm and create unrealistic expectations from the clients. The architect should provide average quality of service that is expected in the area and charge a reasonable fee to do so.)
- Increase the firm's service fees every three years based on CPI and building code updates. (It is not a good idea to increase the firm's service fees every three years based on CPI. It is better to increase the firm's service fees every year based on CPI. An architect needs to train his clients to get used to these fee increases.)
- Try to keep the firm's service fees low, and only increase when necessary. (This is not a good practice. An architect needs to increase the firm's service fees each year based on CPI to keep up with inflation and have enough money to hire or keep good staff who provide quality service.)

23. Answer: a and c
    A **net multiplier** is the ratio of **net revenue** to direct salary expenses. It is also a common way to evaluate staff productivity.

    $$Net\ Multiplier\ =\ \frac{Net\ Revenue}{Direct\ Salary\ Expenses}$$

    Net revenue should exclude consultants' fees, reimbursable expenses, and other "pass-through" costs.

24. Answer: c

    Negligence by the architect is a tort. A **tort** is not a criminal act, but a civil wrongdoing caused by negligence.

25. Answer: e
    A **utilization ratio** is the sum of billable hours divided by the total hours of an employee. It is a measure of the financial well-being and profitability of a firm. Typically, lower level employees have a higher utilization ratio. Higher level employees have a lower utilization ratio because they spend more time in marketing and management of the firm.

26. Answer: b
    A **quick ratio** is the total current assets (account receivable plus cash and equivalent) and unbilled revenue divided by the total of current liabilities. It is a measure of the financial well-being of a firm.

27. Answer: a and b
    The following regarding the joinder provisions in AIA documents are true:
    - They allow parties to consolidate arbitrations.
    - They allow parties include a third party by joinder in arbitrations.

    The following are incorrect answers:
    - They allow parties to join as a team to split legal costs. (This was not in the joinder provisions.)
    - They allow an architect to be the initial decision maker in arbitration. (This was not in the joinder provisions. An architect can be the **initial decision maker** in a dispute resolution, but not in an arbitration. Arbitration is a proceeding handled by an impartial **adjudicator** whose decision will be final and binding. An architect may not qualify to be an adjudicator. Retired judges often work as an adjudicator or **arbitrator.**)

28. Answer: b and e
    The following are true regarding a project alliance:
    - The practice is often used in integrated project delivery (IPD).
    - Project teams make a consensus-based agreement.

The following are incorrect answers:
- Parties waive liabilities against each other. (Liabilities are limited but not completely waived.)
- A project alliance is a single-purpose legal entity, similar to a corporation. (This is *not* true. Integrated project delivery is frequently associated with a project alliance. In IPD, the contractor does not have a contract with the architect. The contractor and architect each have a separate contract with the owner.)
- A project alliance is a form of design-build. (The contractor has a contract with the architect in design-build. This is different from a project alliance.

29. Answer: a and c

    The proper responses by the architect are as follows:
    - Refer her to her own insurance agent or attorney for advice.
    - Give her the AIA document G612, Owner's Instructions to the Architect Regarding the Construction Contract. (The owner can tell the architect which insurances she is carrying, and which insurances she wants the contractor to provide on this form.)

    The following are incorrect answers:
    - Recommend the general liabilities insurance, worker's comp insurance, performance bond, and other typical insurances required for a project and ask her to verify the information herself. (An architect is *not* qualified to make a recommendation for an owner's insurance needs.)
    - Set up a meeting between the owner and the architect's insurance agent. (This might be nice for the architect to do for his friend, but his insurance agent might not carry the type of insurance his friend needs. Determining this requirement is outside of the architect's qualification and scope of work.)

30. Answer: a and d

    The firm will likely make the following changes:
    - Hire new staff as needed.
    - Work with the client to adjust the deadline.

    The following are incorrect answers:
    - Hire new staff quickly to handle the new project. (A new project does not need a big staff at the beginning, so the firm does not need to hire new staff quickly.)
    - Ask the existing staff to work overtime. (This is not a good idea, since overtime pay is expensive, and the staff will not work efficiently if they need to work for a long time.)

31. Answer: b and d

    The best way to minimize risks for an architectural firm are to:
    - Start the effort of minimizing risks at the marketing stage.
    - Select clients with a similar value.

    The following are not the best answers:

- Start the effort of minimizing risks at the feasibility study stage. (This effort should start as soon as possible, like the marketing stage. The architect can start by marking to the right clients.)
- Do research and interview developers. (These tasks will help but are not the best answer.)
- Prepare the contract between the architect and client. (This may help but is not very effective as the client may not agree with the terms the architect proposes.)

32. Answer: d

    A **limited liability company** will minimize the risk for the owner and principals of a firm. Members will *not* have personal liabilities.

    The following are not the best answers:
    - Sole proprietorship will not protect the owner of a firm.
    - Limited partnership requires two or more people, and the firm may not have two or more people who want to take the risks.
    - A joint venture is a temporary organization formed for a specific project.
    - A professional corporation limits the liabilities to the person responsible. Typically, this person is one of the owners or a future principal.

33. Answer: c

    Utilizing studios with experience in the project types is the best organization plan for the expansion.

    The following are not the best answers:
    - Keep the current studio structure. (This will create a mismatch of current staff and the new services.)
    - Create specialty departments. (This may work for a large firm, but is not a good choice for a studio-based architectural firm with only twenty employees.)
    - Convert to a horizontal departmental structure. (This is not a good choice because people with specialty experience like interior design would not be fit to do work for other project types.)

34. Answer: e

    A **design charrette** is the best way to get the client involved in the early stages of design. It is an intense period of design or planning activity that involves all stakeholders of a project.

    The following are not the best answers:
    - phone calls (Some types of information are hard to communicate through phone calls.)
    - questionnaires sent to the client (The client may not fill out the questionnaires on time.)
    - PowerPoint presentations (This is just one way to communicate from the architect to client.)
    - hand sketches (This is typically a one-way type of communication from the architect to client.)

35. Answer: b
Per C401-2007, Standard Form of Agreement Between Architect and Consultant, the electrical engineer should pay for the corrections because he failed to coordinate with the architect and the mechanical engineer's CADD files.

The following are not the best answers:
- architect (The architect has regularly exchanged CADD files with her consultants. The electrical engineer should have coordinated with the architect's plans.)
- The mechanical engineer is not responsible for light locations.
- The general contractor is not responsible because he has no obligation to perform quality control for the architect and her consultants' plans.
- The electrical contractor is not responsible because he followed the electrical plans.

36. Answer: a and c
The most cost-effective way for an eighteen-person firm specializing in retail buildings to create project specifications are to:
- Use an in-house specifications writer to modify the office's prototype specifications. (For an eighteen-person firm specializing in retail buildings, this is cost effective. For a firm with many different types of projects, this will not be a good idea.)
- Hire an outside specifications consultant. (This is cost effective, because the firm does not need to pay taxes or benefits for the consultant.)

The following are not the best answers:
- Subscribe to master specifications and modify them per the specific project. (This is not cost effective because the firm has to pay a monthly fee, and the master specifications are too general for retail projects.)
- Ask the project manager for each project type to develop specifications per the office's prototype specifications. (This is not as cost effective as using an in-house specifications writer.)

37. Answer: a
An architect should most actively involve the elevator consultant regarding vertical transportation. An elevator consultant can help an architect throughout the whole process of the vertical transportation design, while the other consultants will only be involved for a portion of the design.

38. Answer: a
The architect should decline the offer and tell him that the budget is not feasible for what he wants to build.

The following are not the best answers:
- Decline the offer and tell him it is a violation of the AIA *Code of Ethics and Professional Conduct* to accept a project that is clearly under budget. (This is not a violation of the AIA *Code of Ethics and Professional Conduct*.)
- Accept the offer but ask the client to increase the budget. (The client is very unlikely to increase the budget by 30%.)

- Accept the offer and strive to meet the budget in an innovative way. (This is not practical.)

39. Answer: b, c, e, and f

    An architect should know the following about a client to minimize project risks:
    - prior project history (This is a good indicator of potential problems with the client.)
    - financial ability (This is very important.)
    - the client's project team (These people can be helpful or create an obstacle for the project's success.)
    - the client's preferred type of service agreement (If the client wants to use non-AIA documents, there can be a risk factor.)

    The following are not the best answers:
    - personality and character (This should not normally be a deal breaker.)
    - their ideas about how the building should look (This can be discussed and adjusted during the design process.)

40. Answer: c

    Design-assist contracting is a form of project delivery which involves specialty subcontractors or trades early in the design and construction document phase. This practice is often used in the development of unique or complex portions of a building since specialty subcontractors or trades or suppliers may have more knowledge on the special system as compared to the architect or general contractor.

41. Answer: a and c

    The following are true:
    - The architect owns the copyright for the plans of the project. (This is based on B101-2007, Standard Form of Agreement Between Owner and Architect. The owner can only use the plans for this project. If he wants to use the plans for another project, he needs to get the architect's permission, pay the architect a licensing fee, and hold the architect harmless if he does not hire him for the next project.)
    - The architect owns the copyright for the actual building. (The Architectural Works Copyright Protection Act allows an architect to copyright an *actual* building to prevent others from copying his building design.)

42. Answer: c and d

    The following are correct answers:
    - Pay for the correction. (The owner has a contract with the architect. As a prime consultant to the owner, the architect is responsible to the owner for his consultants' work.)
    - Sue the electrical engineer for the cost of the correction plus the legal costs. (Per C401-2007, Standard Form of Agreement Between Architect and Consultant, the consultants are responsible to the architect for their area of work.)

The following are not the best answers:
- Inform the owner that the electrical engineer should pay for it. (The owner does not care about the contract between the architect and his consultants. The owner is entitled to recover his cost from the architect based on his contract with the architect.)
- Send a formal payment request to the electrical engineer and copy the owner in the correspondence. (See above.)

43. Answer: b and c

    ConsensusDOCS are standard contract documents developed by a consortium of about forty construction industry organizations to compete with the AIA documents.

    The following are correct answers:
    - An architect can use ConsensusDOCS.
    - An architect can use AIA documents or ConsensusDOCS.

    The following are not the best answers:
    - An architect should always use AIA documents. (The problem is with the word "always." An architect should always *try* to use AIA documents, but an architect can use non-AIA documents as long as the provisions are reviewed and revised as necessary to protect the architect.)
    - An architect should use ConsensusDOCS. (An architect should always *try* to use AIA documents, and only use ConsensusDOCS if the owner insists on using them.)

44. Answer: e

    If an architect uses an outside drafting consultant, she needs to keep documents of the drafter's involvement and architect's supervision of the work for four to fifteen years, depending on the state she lives in. See "Retaining and Archiving Record" in *the Architect's Handbook of Professional Practice*.

45. Answer: d

    The best way to modify an AIA document is to attach amendments or supplemental conditions.

    The following are not the best answers:
    - Retype the document and then revise. (This is a violation of the copyright of the AIA document.)
    - Cross out the text by hand and write in the changes by hand. (This is OK if the parties involved initial the changes.)
    - Have an attorney draft a new contract using some of the language from the actual AIA document. (This new document will not be familiar to the contractors or architect, and may create more liabilities or increase the project cost.)

46. Answer: d

    Stipulated sum will encourage a contractor to be most efficient.

The following are not the best answers:
- cost-plus with a fixed fee for overhead and profit (The contractor does not care about the construction cost since his own fee is fixed no matter what happens.)
- cost-plus with overhead and profit based on a percentage of the construction cost (The contractor will want to make the construction cost higher since his own fee is based on a percentage of the construction cost.)
- Unit price is only part of the contract when the exact quantities of units cannot be determined.

47. Answer: a

    An architect is liable for any errors and omissions of her consultants. This kind of indirect liability is called **vicarious liability**.

48. Answer: a and b

    The following are typically excluded from an all-risk insurance:
    - act of war
    - employee theft

    The following are normally included in an all-risk insurance:
    - collapse
    - theft
    - debris removal
    - vandalism
    - explosions
    - water damages
    - earthquakes
    - floods
    - windstorms

49. Answer: b and c

    The following are the correct answers:
    - Politely decline the request since it is not cost effective to do the project.
    - Politely decline the request since it is against NCARB's *Model Rules of Conduct* and copyright laws to modify another designer's work.

    The following are not the best answers:
    - Give the client a quote to provide only the missing information needed to obtain the permit. (It is against NCARB's *Model Rules of Conduct* and copyright laws to modify another designer's work.)
    - Send the client a proposal to develop a new set of plans using the standard design as a starting point. (A client who did not even want to pay an architect to design the building to begin with, is unlikely to pay any reasonable fees. It is not cost effective to do the project since the efforts required will exceed the fees that she can charge.)

50. Answer: c
   An architect should keep project records
   • until the end of the statute of repose

   The AIA *Best Practices* advises firms to keep project records for one year past the longest applicable date. Deadlines imposed by the statute of repose are enforced much more strictly than those by the statute of limitations. In many cases, both a statute of limitations and a statute of repose will apply to the same case, and a statute of repose may cut off liability even if the statute of limitations has not run out.

   The following are not the best answers:
   • for four years after substantial construction
   • until the end of the statute of limitations
   • for four years after substantial completion

51. Answer: a
   An owner uses design development plans to obtain a fixed price bid, and then allows the contractor to decide the details of the construction. This method of construction delivery is called **bridging**. The owner's architect develops the design development plans, and the contractor has his own in-house architect to develop the construction plans based on the design development plans provided by the owner.

   The following are not the best answers:
   • **design-build** (**Bridging** is a special type of design-build.)
   • **Design-bid-build** is the traditional delivery method.
   • **The definition of integrated project delivery by Wikipedia** "is a collaborative alliance of people, systems, business structures, and practices into a process that harnesses the talents and insights of all participants to optimize project results, increase value to the owner, reduce waste, and maximize efficiency through all phases of design, fabrication, and construction."
   • **Fast-track** allows construction to start before the plans are fully completed.

52. Answer: b and c
   The following are true regarding the **construction management at risk (CMAR)** project delivery method:
   • It has two separate contracts. (The construction manager typically takes on the risk of performance, signs a guaranteed maximum price contract offer with the owner before the construction documents are completed, and then subcontracts the work to other contractors).
   • It has three prime players (owner, architect, and contractor).

   The following are not the correct answers:
   • The construction documents must be completed before construction starts.
   • It has three prime contractors. (It has three prime *players*, *not* three prime *contractors*.)

79 • Chapter Four

53. Answer: c

If a supermarket owner wants to make sure the market can have a grand opening within one year, the best project delivery method is fast track. Fast track allows the design and construction phases to overlap in order to complete a project faster.

54. Answer: a and d

The best courses of action for the architect are to:
- Ask the client to pay additional fees (A request to reduce the time needed for design and construction documents after the contract is signed will mean extra cost to the architect, and the architect will need to pass this cost back to the client.)
- Ask the client to reduce the scope of the project. (This is another option. A reduced project scope will reduce the time needed to complete.)

The following are not the correct answers:
- Hire new employees to speed up the project. (This strategy will incur extra costs and should not be done unless the client has approved the architect's request for more money.)
- Assign more existing staff to the project (This will impact other projects.)
- Ask the project team to work overtime (This strategy will incur extra costs and should not be done unless the client has approved the architect's request for more money.)

55. Answer: c

The best course of action for the architect is to:
- Send a written notice asking the owner to cease and desist.

B101, Standard Form of Agreement Between Owner and Architect grants the owner a non-exclusive license to use the instrument of service (including construction documents) to construct the project. When the contract is terminated for non-payment of the service, this non-exclusive license is terminated too, and the owner no longer has the right to use the construction documents.

The following are not the correct answers:
- Do nothing.
- Send a written notice asking the owner to pay for the services rendered and informing him that the architect will take legal action if not reimbursed promptly. (The contract was already terminated for non-payment of services rendered. The architect must have sent out the request for payment to the client many times already to no avail)
- none of the above

56. Answer: c

According to the **MacLeamy curve**, an architect should allocate the most resources during the detailed design phase of the project.

The integrated project delivery (IPD) method front loads the efforts to the early stages of a project. The greatest effort occurs at design development phase or detailed design phase.

The follow is a comparison of the integrated project delivery (IPD) and traditional delivery method:

Integrated Project Delivery (IPD)	Traditional Delivery Method
a. conceptualization	a. predesign
b. criteria design	b. schematic design
c. detailed design	c. design development
d. implementation document	d. construction documents
e. agency coordination	e. bidding
f. construction	f. construction

Figure 4.2   A Comparison of Integrated Project Delivery (IPD) and Traditional Delivery Method

57. Answer: c
They should use a teaming agreement.
A joint venture agreement or prime consulting agreement should be used after they win the bid and are awarded the project. Corporation agreement is just a distractor.

58. Answer: a, b, c, and d

The following are mandatory requirements for NCARB's **Architectural Experience Program (AXP)**:
- a licensed architect as supervisor
- a NCARB record
- experience in specific categories
- online reporting

State coordinators and interactive online service are helpful, but they are not mandatory.

59. Answer: b
The biggest impact on the firm is the need to give a reason for terminating an employee if the employment is no longer at will.

After the change, the book keeping requirements pretty much stay the same. Depending on the employment contract, moonlighting may or may not be prohibited. Specific job descriptions for each position may not be required.

60. Answer: d
A twelve-person architectural firm should use a **modified accrual-basis method**.

The **accrual-basis method** keeps track of expenses and revenues when they are incurred.
A **cash-basis method** is used by very small firms or sole proprietors and is not proper for this situation. Revenue and expenses are recognized when transactions are made.
The **double-entry method** keeps track of transactions with the accrual-basis method.

A **modified accrual-basis method** includes the advantages of both the accrual-basis method and cash-basis method.

61. Answer: a

Cost-plus fee is the best pricing method. This method will make sure that the costs are covered, and the architect and client can reach a fair fee agreement.

Percentage of construction cost is not good because the client may think the architect will increase the construction cost to get a higher fee. If the project is small, the architect will not be paid fairly.

Hourly not to exceed is not the best method because the client is inexperienced, and his indecision and changes can easily cause the fee to go well beyond the limit. If the architect sets a higher limit, the client may feel uncomfortable.

Square-feet cost uses historical data and may not cover the changes from an inexperienced client.

62. Answer: b, e, and f

The normal **overhead rate** should be 1.3 to 1.5. The architect should do the following to reduce the overhead:
- Monitor unbillable hours more closely. (This can help reduce the unbillable hours.)
- Move to a cheaper office. (This can generate significant savings.)
- Ask the administrative staff to check their direct expenses. (Project-related hours of administrative staff should be billed to the project.)

The following are not the best answers:
- Review reimbursables. (This has no impact since reimbursables will be passed through to the clients.)
- Be vigilant in fee collection. (This can help the architect to get the fee more quickly, but it cannot reduce overhead.)
- Increase revenue per technical staff. (This will help increase the profit but will not reduce overhead.)

63. Answer: a

An architect should use the **profit and loss statement** to evaluate the financial performance of her firm since it gives an overall view of the profit and loss.

The following are not the best answers:
- Net profit before tax shows the percentage of the net revenue excluding reimbursables and consulting fees.
- The balance sheet shows the assets, liabilities, and net worth of a firm.
- The office earning report focuses on the financial performance of individual projects, not the overall performance of the firm.

64. Answer: c and d
    The following are proper strategies:
    - Research the market and identify possible contacts. (This is a good start.)
    - Hire a professional with experience in retail buildings and office buildings. (This is a good strategy, because the professional can help to research the retail market and bring in key contacts. When his workload is light, he can help with the office building projects.)

    The following are not the best answers:
    - Develop a web page on the firm's website showcasing retail building design. (This strategy is too early as the firm has no portfolio in this field to show.)
    - Start social media marketing. (This should be started after the firm has researched the retail market.)

65. Answer: a, c e, and f
    The following are most effective for **quality control** during construction document production:
    - Assign one person to be in charge of the project. (This can help avoid confusion, and reduce errors and omissions.)
    - Exchange progress plans with the consultants regularly. (This is a good way to encourage communication and coordination.)
    - Use office standard details. (This will help to reduce error and maintain a certain level of professionalism.)
    - Use CSI's Uniform Drawing System (UDS) of National CAD Standard (NCS). (This is an industry standard.)

    The following are not the best answers:
    - Hire an outside firm to do final quality control. (A better way is to get an experienced in-house architect who has not worked on the project to take a fresh look at the project.)
    - Use the construction specific institute (CSI) checklist. (CSI does not have a checklist.)

B. Mock Exam Answers and Explanations: Case Study

66. Answer: b
    Per B101, Standard Form of Agreement Between Owner and Architect, if the insurance requirements set forth in B101 exceed the types and limits the architect normally maintains, the owner shall reimburse the architect for any additional costs.

67. Answer: c
    A **joint venture** is typically formed when a firm is pursuing a project and does not have the ability to do the job alone. In this case, the firm has already won the commission, the in-house staff members of the firm have the abilities to do the project, but they just do not have enough time.

    An **outside freelance consultant** may have many other projects, and not be as committed as another consulting firm or an employee.

    A **teaming agreement** is an informal collaboration between firms for marketing efforts. It is not a formal business organization.

    A **prime contract agreement** will allow the firm to use the staff and skills of another firm but retain the role of the prime contact with the client and architect of the record. It is the best choice for this situation.

68. Answer: d
    Per B101, Standard Form of Agreement Between Owner and Architect, the architect's cost estimates represent the best professional judgement. However, the architect cannot guarantee that the actual costs will not exceed the cost estimates or project's budget. By agreeing to the owner's request to guarantee the project will not be over the budget, the architect has raised the standard of care.

    However, doing so does not violate the anti-trust laws, void his error and omission professional liabilities insurance, or violate the AIA *Code of Ethics and Professional Conduct*.

69. Answer: b, c, d, and e

    The following types of consultants should be retained to complete the fast-food restaurant projects:
    - structural
    - mechanical
    - electrical
    - plumbing

Since the client provided the prototype plans, the following consultants are not necessary:
- fixture
- signage
- interior designer
- egress
- security

70. Answer: d

If there are some major discrepancies between the mechanical and electrical plans, the architect should try his best to have his consultants correct the plans. The bid is due in one week, so there is not enough time to correct the plans, reissue them to the bidders, as well as give the bidders plenty of time to review the changes. The architect should inform the owner of the situation and ask the owner if he can postpone the bid so that the numbers can be as accurate as possible.

The following are not the best answers:
- Finish the bidding process and ask the engineers to coordinate and correct the plans after the bidding. (This is a solution, but not the best answer since the bid will not be as accurate.)
- Ask the engineers to coordinate and correct the plans and issue the corrected plans to the bidder who brought up the issue. (Any changes *have to* be issued to *all* bidders to be fair.)
- Ask the engineers to coordinate and correct the plans and issue the corrected plans to all bidders. (There is not enough time to do this.)

71. Answer: c

The contractor should bear the cost of replacing the damaged plywood sheathing because he is responsible for protecting the materials at the job site. The architect has done his due diligence and already informed the owner and contractor of this issue.

72. Answer: c

The legal concept that can protect the architect from the claim by the mechanical contractor is **lack of privity.** There is no contract between the architect and mechanical subcontractor.

Agency means someone is representing another for one or more tasks.

Fiduciary duty means a legal duty of one party to act in the best interest of another, such as a defense attorney's duty to a defendant.

Indemnity is a contractual duty of one party (indemnifier) to reimburse the loss occurred to the other party (indemnity holder) due to the act of the indemnitor or someone else.

73. Answer: d
Developing a project check list may be beneficial to the firm. This is a good way to ensure past mistakes will not be repeated.

Asking one of the principals to do quality control of the documents is not practical because the principal for a firm this size will not have time to do this.

Asking the project manager to be more careful and spend more time checking the documents is not much help. The project manager needs to focus on coordinating and passing the information between the designer, the production staff, and other consultants.

Hiring an outside firm to check the documents will incur extra costs.

74. Answer: d
Design the buildings according to the client's prototype, submit them for the cities' review, and apply for variances as needed.

Many cities will approve variances, and this option will make sure the buildings look as close to the client's prototype as possible. Since this option may involve more time and effort, the architect may want to issue an additional service request asking for more fees to cover his costs.

The following are not the best answers:
- Design the buildings according to the client's prototypes. (If the prototypes do not comply with cities' planning guidelines, the plans may not be approved, and the buildings may not be built.)
- Design the buildings according to the cities' planning guidelines. (This option does not address the client's requirements.)
- Design the buildings according to both the client's prototypes and the cities' planning guidelines (This solution does not take advantage of variances that may allow the client to keep more features of his prototype.)

75. Answer: b
Based on the program, the architectural firm has twenty-three employees.
A **single studio** requires experienced staff capable of handling projects from beginning to end and is not the most cost-efficient way to organize this office. There may not be enough work to keep a single studio busy, let alone **two specialist studios**.

A strict **departmental organization** is more appropriate for larger firms. **A department with two design specialists** is the best choice since the specialists will have the expertise to design the projects per their unique requirements, and move the project through different departments such as design, production, specifications, and construction administration.

76. Answer: d
The owner's insurance company should pay for the fire damages of the burnt down building. The owner has the duty to maintain property insurance on an "all risk" or equivalent policy form. This type should include coverage for fire, earthquake, collapse, etc. See A201, General Conditions of the Contract for Construction, for more information.

77. Answer: a
The owner should pay for the extra field visits because they are above and beyond the scope of the original service agreement.

The structural engineer is typically under contract by the architect, so the architect needs to prepare an additional service request, send it to the owner, and obtain the owner's written approval before sending the structural engineers into the field.

78. Answer: e
The contractor's insurance company should pay for this accident.

The contractor shall maintain insurance covering his direct employees and those of the subcontractors. The insurance covers worker's compensation, bodily injuries, death, etc. See A201, General Conditions of the Contract for Construction, for more information.

79. Answer: a and e
The owner should get the approval from the city and owner's insurance company before he moves in and takes partial occupancy. The city may or may not approve the owner's request for partial occupancy. The contractor shall assist the owner in getting approval from the owner's insurance company. See A201, General Conditions of the Contract for Construction, for more information.

80. Answer: a and c
The following LEED certifications are most likely to be achieved:
- LEED BD+C (Because the project is at the end of the construction, it may be too late to implement many of the LEED points for this certification. LEED BD+C is less likely to achieve certification as compared to LEED O+M, but is still more likely than the other choices.)
- LEED O+M (This is most likely to be achieved since LEED O+M typically starts after construction is completed.)

The following are not the best answers:
- LEED ID+C (Retail shop buildings typically only have a completed building shell and no interior construction. The interior design and construction will be done later after each space is leased out. LEED ID+C is typically achieved by individual tenants of each space. The building owner is also the landlord in this case, and it is unlikely for him to achieve LEED ID+C certification.)
- LEED Retail (Retail is a subcategory of the system, such as LEED BD+C: Retail or LEED O+M: Retail.)

The USGBC has the following reference guides and green rating system portfolios.

1) *The LEED Reference Guide for Building Design and Construction v4 (BD&C)* covers the following LEED rating systems:
   - LEED BD+C: New Construction
   - LEED BD+C: Core and Shell Development
   - LEED BD+C: Schools
   - LEED BD+C: Retail
   - LEED BD+C: Data Center
   - LEED BD+C: Warehouse and Distribution Center
   - LEED BD+C: Hospitality
   - LEED BD+C: Health Care
   - LEED BD+C: Homes
   - LEED BD+C: Multi-family Midrise

2) *The LEED Reference Guide for Interior Design and Construction v4 (ID&C)* covers the following LEED rating systems:
   - LEED ID+C: Commercial Interiors
   - LEED ID+C: Retail
   - LEED ID+C: Hospitality

3) *The LEED Reference Guide for Green Building Operations and Maintenance v4 (O&M)* covers the following LEED rating systems:
   - LEED O+M: Existing Buildings
   - LEED O+M: Retail
   - LEED O+M: Schools
   - LEED O+M: Hospitality
   - LEED O+M: Data Center
   - LEED O+M: Warehouse and Distribution Center

   Note: These LEED-O+M rating systems are the only systems that cover building operation. All other LEED systems cover building design and construction, but NOT operation.

4) *The LEED for Neighborhood Development Reference Guide v4 (LEED-ND)* covers the following:
   - LEED ND: Plan
   - LEED ND: Built Project

C. How We Came Up with the Practice Management (PcM) Mock Exam Questions

We came up with all the CE Mock Exam questions based on the ARE 5.0 Handbook, and we developed the Mock Exam based on the *four* weighted sections. See a detailed breakdown in the following tables:

*Note: If the text on following tables is too small for you to read, then you can go to our forum, sign up for a free account, and download the FREE full-size jpeg format files for these tables at **GeeForum.com***

Sections	Expected Number of Items	Actual Number of Items
Total	80	80
Section 1: Business Operations (20-26%)	16-21	19
• Assess resources within the practice (A/E)		3
• Apply the regulations and requirements governing the work environment (U/A)		5
• Apply ethical standards to comply with accepted principles within a given situation (U/A)		4
• Apply appropriate Standard of Care within a given situation (U/A)		7
Section 2: Finances, Risk, & Development of Practice (29-35%)	23-28	26
• Evaluate the financial well-being of the practice (A/E)		7
• Identify practice policies and methodologies for risk, legal exposures, and resolutions (U/A)		9
• Select and apply practice strategies for a given business situation and policy (U/A)		10
Section 3: Practice-Wide Delivery of Services (22-28%)	17-23	20
• Analyze and determine response for client services requests (A/E)		6
• Analyze applicability of contract types and delivery methods (A/E)		9
• Determine potential risk and/or reward of a project and its impact on the practice (A/E)		5
Section 4: Practice Methodologies (17-23%)	13-18	15
• Analyze the impact of practice methodologies relative to structure and organization of the practice (A/E)		6
• Evaluate design, coordination, and documentation methodologies for the practice (A/E)		9

Figure 4.3 How We Came Up with the Practice Management (PcM) Mock Exam Questions

Appendixes

A. List of Figures

Figure 1.1 The relationship between ARE 4.0 and ARE 5.0..18

Figure 1.2 The hours required under the six experience areas..19

Figure 1.3 Exam format & time...24

Figure 1.4 New Exam format & time...24

Figure 2.1 Exam Content...37

Figure 2.2 Mnemonics for the 2004 CSI MasterFormat..41

Figure 3.1 Employee Workload..44

Figure 3.2 Architect Frank Work Hours...45

Figure 4.1 Architect Frank Work Hours...64

Figure 4.2 A Comparison of Integrated Project Delivery (IPD) and Traditional Delivery Method..80

Figure 4.3 How We Came Up with the Practice Management (PcM) Mock Exam Questions..88

B. Official reference materials suggested by NCARB

1. Resources Available While Testing
Tips:
- *You need to read through these pages several times and become very familiar with them to save time in the real ARE exams.*

United States. American Institute of Steel Construction, Inc. *Steel Construction Manual*; 14th edition. Chicago, Illinois, 2011.

Beam Diagrams and Formulas:
- Simple Beam: Diagrams and Formulas - Conditions 1-3, page 3-213; Conditions 4-6, page 3-214; Conditions 7-9, page 3-215
- Beam Fixed at Both Ends: Diagrams and Formulas - Conditions 15-17, page 3-218
- Beam Overhanging One Support: Diagrams and Formulas - Conditions 24-28, pages 3-221 & 222

Dimensions and Properties:
- W Shapes 44 thru 27: Dimensions and Properties, pages 1-12 thru 17
- W Shapes 24 thru W14x145: Dimensions and Properties, pages 1-18 thru 23
- W Shapes 14x132 thru W4: Dimensions and Properties, pages 1-24 thru 29
- C Shapes: Dimensions and Properties, pages 1-36 & 37
- Angles: Properties, pages 1-42 thru 49
- Rectangular HSS: Dimensions and Properties, pages 1-74 thru 91
- Square HSS: Dimensions and Properties, pages 1-92 thru 95
- Round HSS: Dimensions and Properties, pages 1-96 thru 100

United States. International Code Council, Inc. *2012 International Building Code.* Country Club Hills, Illinois, 2011.

Live and Concentrated Loads:
- Uniform and Concentrated Loads: IBC Table 1607.1, pages 340-341

2. Typical Beam Nomenclature

The following typical beam nomenclature is excerpted from:
United States. American Institute of Steel Construction, Inc. *Steel Construction Manual*; 14th edition. Chicago, Illinois, 2011.

E	Modulus of Elasticity of steel at 29,000 ksi	V_2	Vertical shear at right reaction point, or to left of intermediate reaction of beam, kips
I	Moment of Inertia of beam, in^4	V_3	Vertical shear at right reaction point, or to right of intermediate reaction of beam, kips
L	Total length of beam between reaction point, ft	V_x	Vertical shear at distance x from end of beam, kips
M_{max}	Maximum moment, kip-in	W	Total load on beam, kips
M_1	Maximum moment in left section of beam, kip-in	A	Measured distance along beam, in
M_2	Maximum moment in right section of beam, kip-in	B	Measured distance along beam which may be greater or less than a, in
M_3	Maximum positive moment in beam with combined end moment conditions, kip-in	L	Total length of beam between reaction points, in
M_x	Maximum at distance x from end of beam, kip-in	w	Uniformly distributed load per unit of length, kips/in
P	Concentrated load, kips	w_1	Uniformly distributed load per unit of length nearest left reaction, kips/in
P_1	Concentrated load nearest left reaction, kips	w_2	Uniformly distributed load per unit of length nearest right reaction and of different magnitude than w1, kips/in
P_2	Concentrated load nearest right reaction and of different magnitude than P_1, kips	X	Any distance measured along beam from left reaction, in
R	End beam reaction for any condition of symmetrical loading, kips	x_1	Any distance measured along overhang section of beam from nearest reaction point, in
R_1	Left end beam reaction, kips	Δ_{max}	Maximum deflection, in
R_2	Right end or intermediate beam reaction, kips	Δa	Deflection at point of load, in
R_3	Right end beam reaction, kips	Δx	Deflection at point x distance from left reaction, in
V	Maximum vertical shear for any condition of symmetrical loading, kips	Δx_1	Deflection of overhang section of beam at any distance from nearest reaction point, in
V_1	Maximum vertical shear in left section of beam, kips		

3. Formulas Available While Testing

Tips:
- *These formulas and references will be available during the real exam. You should read through them a few times before the exam to become familiar with them. This will save you a lot of time during the real exam, and will help you solve structural calculations and other problems.*

Structural:
Flexural stress at extreme fiber
$$f = \frac{Mc}{I} = \frac{M}{S}$$

Flexural stress at any fiber
$$f = \frac{My}{I}$$

where y = distance from neutral axis to fiber

Average vertical shear
$$v = \frac{V}{A} = \frac{V}{dt}$$
for beams and girders

Horizontal shearing stress at any section A-A
$$v = \frac{VQ}{Ib}$$
where Q = statical moment about the neutral axis of the entire section of that portion of the cross-section lying outside of section A-A
b = width at section A-A

Electrical
$$Foot - candles = \frac{lumens}{area\ in\ ft^2}$$

$$Foot - candles = \frac{(lamp\ lumens)\ x\ (lamps\ per\ fixture)\ x\ (number\ of\ fixtures)\ x\ (CU)\ x\ (LLF)}{area\ in\ ft^2}$$

$$Number\ of\ luminaires = \frac{(foot - candles)\ x\ (floor\ area)}{(lumens)\ x\ (CU)\ x\ (LLF)}$$

where CU = coefficient of utilization
LLF = Light Loss Factor

$$DF_{AV} = 0.2x \frac{window\ area}{floor\ area}$$
for spaces with side lighting or top lighting with vertical monitors

watts = volts x amperes x power factor
for AC circuits only

Demand charge = maximum power demand x demand tariff

Plumbing
1 psi = 2.31 feet of water

1 cubic foot = 7.5 U.S. gallons

HVAC

$$\frac{BTU}{year} = peak\ heat\ loss \times \frac{full - load\ hours}{year}$$

$$BTU/h = (cfm)\ x\ (1.08)\ x\ (\Delta T)$$

1 kWh = 3,400 BTU/h

1 ton of air conditioning = 12,000 BTU/h

$BTU/h = (U)\ x\ (A)\ x\ (T_d)$ *where Td is the difference between indoor and outdoor temperatures*

$$U = 1/R_t$$

$$U_o = \frac{(U_w \times A_w) + (U_{op} \times A_{op})}{Ao}$$
where o = total wall, w = window, and op = opaque wall

$$U_o = \frac{(U_R \times A_R) + (U_S \times A_S)}{A_o}$$
where o = total roof, R = roof, and S = skylight

$$R = x/k$$
where x = thickness of material in inches

$$\text{Heat required} = \frac{BTU/h}{\text{temperature differential}} \times (24 \text{ hours}) \times (DD \, °F)$$
where DD = degree days

Acoustics
$$\lambda = \frac{c}{f}$$
where λ = wavelength of sound (ft)
c = velocity of sound (fps)
f = frequency of sound (Hz)

$$a = SAC \times S$$
where a = Absorption of a material used in space (sabins)
SAC = Sound Absorption Coefficient of the material
S = Exposed surface area of the material (ft^2)

$$A = \Sigma a$$
Where A = Total sound absorption of a room (sabins)
$\Sigma a = (S_1 \times SAC_1) + (S_2 \times SAC_2) + ...$

$$T = 0.05 \times \frac{V}{A}$$
where T = Reverberation time (seconds)
V = Volume of space (ft^3)

$$NRC = \text{average SAC for frequency bands } 250, 500, 1000, \text{ and } 2000 \, Hz$$

4. Common Abbreviations

Tips:
- *You need to read through these common abbreviations several times and become very familiar with them to save time in the real ARE exams.*

Professional Organizations, Societies, and Agencies

American Concrete Institute	ACI
American Institute of Architects	AIA
American Institute of Steel Construction	AISC
American National Standards Institute	ANSI
American Society for Testing and Materials	ASTM
American Society of Civil Engineers	ASCE
American Society of Heating, Refrigerating, and Air-Conditioning Engineers	ASHRAE
American Society of Mechanical Engineers	ASME
American Society of Plumbing Engineers	ASPE
Architectural Woodwork Institute	AWI
Construction Specifications Institute	CSI
Department of Housing and Urban Development	HUD
Environmental Protection Agency	EPA
Federal Emergency Management Agency	FEMA
National Fire Protection Association	NFPA
Occupational Safety and Health Administration	OSHA
U.S. Green Building Council	USGBC

Tips:
- *You need to look through the following codes and regulations & AIA contract documents several times and become very familiar with them to save time in the real ARE exams. Read some of the important sections in details.*

AIA Contract Documents

A101-2007, Standard Form of Agreement Between Owner and Contractor - Stipulated Sum	A101
A201-2007, General Conditions of the Contract for Construction	A201
A305-1986, Contractor's Qualification Statement	A305
A701-1997, Instructions to Bidders	A701
B101-2007, Standard Form of Agreement Between Owner and Architect	B101
C401-2007, Standard Form of Agreement Between Architect and Consultant	C401
G701-2001, Change Order	G701
G702-1992, Application and Certificate for Payment	G702
G703-1992, Continuation Sheet	G703
G704-2000, Certificate of Substantial Completion	G704

Codes and Regulations

ADA Standards for Accessible Design	ADA

International Code Council	ICC
International Building Code	IBC
International Energy Conservation Code	IECC
International Existing Building Code	IEBC
International Mechanical Code	IMC
International Plumbing Code	IPC
International Residential Code	IRC
Leadership in Energy and Environmental Design	LEED
National Electrical Code	NEC

Commonly Used Terms

Air Handling Unit	AHU
Authority Having Jurisdiction	AHJ
Building Information Modeling	BIM
Concrete Masonry Unit	CMU
Contract Administration	CA
Construction Document	CD
Dead Load	DL
Design Development	DD
Exterior Insulation and Finish System	EIFS
Furniture, Furnishings & Equipment	FF&E
Floor Area Ratio	FAR
Heating, Ventilating, and Air Conditioning	HVAC
Insulating Glass Unit	IGU
Indoor Air Quality	IAQ
Indoor Environmental Quality	IEQ
Live Load	LL
Material Safety Data Sheets	MSDS
Photovoltaic	PV
Reflected Ceiling Plan	RCP
Schematic Design	SD
Variable Air Volume	VAV
Volatile Organic Compound	VOC
British Thermal Unit	btu
Cubic Feet per Minute	cfm
Cubic Feet per Second	cfs
Cubic Foot	cu. ft. ft^3
Cubic Inch	cu. in. in^3
Cubic Yard	cu. yd. yd^3
Decibel	dB
Foot	ft
Foot-candle	fc
Gross Square Feet	gsf

Impact Insulation Class	IIC
Inch	in
Net Square Feet	nsf
Noise Reduction Coefficient	NRC
Pound	lb
Pounds per Linear Foot	plf
Pounds per Square Foot	psf
Pounds per Square Inch	psi
Sound Transmission Class	STC
Square Foot	sq. ft.
	sf
	ft^2
Square Inch	sq. in.
	in^2
Square Yard	sq. yd.

5. General NCARB reference materials for ARE:

Per NCARB, all candidates should become familiar with the latest version of the following codes:

International Code Council, Inc. (ICC)
International Building Code
International Mechanical Code
International Plumbing Code

National Fire Protection Association (NFPA)
Life Safety Code (NFPA 101)
National Electrical Code (NFPA 70)

National Research Council of Canada
National Building Code of Canada
National Plumbing Code of Canada
National Fire Code of Canada

American Institute of Architects
AIA Documents - 2007

6. Official NCARB reference materials matrix

Per NCARB, all candidates should become familiar with the latest version of the following:

Reference	PcM	PjM	PA	PPD	PDD	CE
ADA Standards for Accessible Design U.S. Department of Justice, Latest Edition			X	X	X	
Code of Ethics and Professional Conduct AIA Office of General Counsel. The American Institute of Architects, latest edition	X					
The Architect's Handbook of Professional Practice The American Institute of Architects John Wiley & Sons, latest edition	X	X	X	X	X	X
The Architect's Studio Companion: Rules of Thumb for Preliminary Design Edward Allen and Joseph Iano John Wiley & Sons, 6th edition, 2017			X	X		
Architectural Acoustics. M. David Egan. J. Ross Publishing. Reprint. Original publication McGraw Hill, latest edition			X	X	X	
Architectural Graphic Standards The American Institute of Architects John Wiley & Sons, latest edition			X	X	X	
Building Codes Illustrated: A Guide to Understanding the International Building Code. Francis D.K. Ching and Steven R. Winkel, FAIA, PE. John Wiley & Sons, latest edition			X	X	X	
Building Construction Illustrated Francis D. K. Ching John Wiley & Sons, latest edition				X	X	
Building Structures James Ambrose and Patrick Tripeny John Wiley & Sons, 3rd edition, Latest Edition			X	X	X	
CSI MasterFormat. The Construction Specifications Institute, latest edition					X	X
Daylighting Handbook I Christoph Reinhart Building Technology Press, latest edition			X	X		
Dictionary of Architecture and Construction. Cyril M. Harris. McGraw-Hill, Latest edition			X	X	X	
Framework for Design Excellence American Institute of Architects Available Online			X	X		

Reference	PcM	PjM	PA	PPD	PDD	CE
Fundamentals of Building Construction: Materials and Methods Edward Allen and Joseph Iano John Wiley & Sons, latest edition				■	■	
Green Building Illustrated Francis D.K. Ching and Ian M. Shapiro Wiley, latest edition				■	■	
The Green Studio Handbook: Environmental Strategies for Schematic Design Alison G. Kwok and Walter Grondzik Routledge, latest edition			■	■		
Heating, Cooling, Lighting: Sustainable Design Methods for Architects. Norbert Lechner. John Wiley & Sons, latest edition				■	■	
The HOK Guidebook to Sustainable Design Sandra F. Mendler, William Odell, and Mary Ann Lazarus John Wiley & Sons, latest edition			■	■	■	
ICC A117.1-2009 Accessible and Usable Buildings and Facilities International Code Council, 2010			■	■	■	
International Building Code International Code Council, latest edition			■	■	■	
Law for Architects: What You Need to Know. Robert F. Herrmann and the Attorneys at Menaker & Herrmann LLP. W. W. Norton, latest edition	■					
Legislative Guidelines and Model Law/Model Regulations National Council of Architectural Registration Boards, latest edition	■					
Mechanical & Electrical Equipment for Buildings. Walter T. Grondzik, Alison G. Kwok, Benjamin Stein, and John S. Reynolds, Editors. John Wiley & Sons, latest edition				■	■	
Mechanical and Electrical Systems in Buildings. Richard R. Janis and William K. Y. Tao. Prentice Hall, latest edition				■	■	
Model Rules of Conduct National Council of Architectural Registration Boards, latest edition	■					

Reference	PcM	PjM	PA	PPD	PDD	CE
Olin's Construction Principles, Materials, and Methods. H. Leslie Simmons. John Wiley & Sons, latest edition				■	■	
Planning and Urban Design Standards American Planning Association John Wiley & Sons, latest edition			■	■		
Plumbing, Electricity, Acoustics: Sustainable Design Methods for Architecture. Norbert Lechner. John Wiley & Sons, latest edition				■	■	
Problem Seeking: An Architectural Programming Primer William M. Peña and Steven A. Parshall John Wiley & Sons, latest edition			■			
Professional Practice: A Guide to Turning Designs into Buildings. Paul Segal, FAIA. W. W. Norton, latest edition	■	■	■			
The Professional Practice of Architectural Working Drawings. Osamu A. Wakita, Nagy R. Bakhoum, and Richard M. Linde. John Wiley & Sons, latest edition			■	■	■	
The Project Resource Manual: CSI Manual of Practice. The Construction Specifications Institute. McGraw-Hill, latest edition		■			■	■
Simplified Engineering for Architects and Builders James Ambrose and Patrick Tripeny John Wiley & Sons, latest edition				■	■	
Site Planning and Design Handbook Thomas H. Russ McGraw-Hill, latest edition			■			
Space Planning Basics Mark Karlen and Rob Fleming John Wiley & Sons, latest edition			■			
Steel Construction Manual American Institute of Steel Construction Ingram, latest edition					■	
Structural Design: A Practical Guide for Architects James R. Underwood and Michele Chiuini John Wiley & Sons, latest edition				■	■	
Structures Daniel Schodek and Martin Bechthold Pearson/Prentice Hall, latest edition						

Reference	PcM	PjM	PA	PPD	PDD	CE
Sun, Wind, and Light: Architectural Design Strategies G.Z. Brown and Mark DeKay John Wiley & Sons, latest edition			■	■		
Sustainable Construction: Green Building Design and Delivery Charles J. Kibert. John Wiley & Sons, latest edition			■	■		
A Visual Dictionary of Architecture Francis D.K. Ching John Wiley & Sons, latest edition				■	■	

The following AIA Contract Documents have content covered in the of ARE 5.0 exams. Candidates can access them for free through their NCARB Record.

Document	PcM	PjM	PA	PPD	PDD	CE
A101-2017, Standard Form of Agreement Between Owner and Contractor where the basis of payment is a Stipulated Sum		■				■
A133-2019, Standard Form of Agreement Between Owner and Construction Manager as Constructor where the basis of payment is the Cost of the Work Plus a Fee with a Guaranteed Maximum Price		■				■
A195-2008, Standard Form of Agreement Between Owner and Contractor for Integrated Project Delivery		■				
A201-2017, General Conditions of the Contract for Construction		■				■
A295-2008, General Conditions of the Contract for Integrated Project Delivery		■				
A305-1986, Contractor's Qualification Statement						■
A701-2018, Instructions to Bidders						■
B101-2017, Standard Form of Agreement Between Owner and Architect	■	■				■
B195-2008, Standard Form of Agreement Between Owner and Architect for Integrated Project Delivery		■				
C401-2017, Standard Form of Agreement Between Architect and Consultant	■	■				■

Document	PcM	PjM	PA	PPD	PDD	CE
G701-2017, Change Order						
G702-1992, Application and Certificate for Payment						
G703-1992, Continuation Sheet						
G704-2017, Certificate of Substantial Completion						

The following are some extra study materials if you have some additional time and want to learn more. If you are tight on time, you can simply look through them and focus on the sections that cover your weakness:

ACI Code 318-05 (Building Code Requirements for Reinforced Concrete)
American Concrete Institute, 2005

OR
CAN/CSA-A23.1-94 (Concrete Materials and Methods of Concrete Construction) and CAN/CSA-A23.3-94 (Design of Concrete Structures for Buildings)
Canadian Standards Association

Design Value for Wood Construction
American Wood Council, 2005

Elementary Structures for Architects and Builders, Fourth Edition
Ronald E. Shaeffer
Prentice Hall, 2006

Introduction to Wood Design
Canadian Wood Council, 2005

Manual of Steel Construction: Allowable Stress Design; 9th Edition.
American Institute of Steel Construction, Inc. Chicago, Illinois, 1989

National Building Code of Canada, 2005
Parts 1, 3, 4, 9; Appendix A
Supplement
Chapters 1, 2, 4; Commentaries A, D, F, H, I

NEHRP (National Earthquake Hazards Reduction Program) Recommended Provisions for Seismic Regulations for New Buildings and Other Structures Parts 1 and 2
FEMA 2003

Simplified Building Design for Wind and Earthquake Forces

James Ambrose and Dimitry Vergun
John Wiley & Sons, 1997

Simplified Design of Concrete Structures,
Eighth Edition
James Ambrose, Patrick Tripeny
John Wiley & Sons, 2007

Simplified Design of Masonry Structures
James Ambrose
John Wiley & Sons, 1997
Simplified Design of Steel Structures, Eighth Edition
James Ambrose, Patrick Tripeny
John Wiley & Sons, 2007

Simplified Design of Wood Structures, Fifth Edition
James Ambrose
John Wiley & Sons, 2009

Simplified Mechanics and Strength of Materials, Fifth Edition
Harry Parker and James Ambrose
John Wiley & Sons, 2002

Standard Specifications Load Tables &Weight Tables for Steel Joists and Joist Girders
Steel Joist Institute, latest edition

Steel Construction Manual, Latest edition
American Institute of Steel Construction, 2006

OR
Handbook of Steel Construction, Latest edition; and *CAN/CSA-S16-01 and CISC Commentary*
Canadian Institute of Steel Construction

Steel Deck Institute Tables
Steel Deck Institute

OR
LSD Steel Deck Tables
Caradon Metal Building Products

Structural Concepts and Systems for Architects and Engineers, Second Edition
T.Y. Lin and Sidney D. Stotesbury
Van Nostrand Reinhold, 1988

Structural Design: A Practical Guide for Architects

James Underwood and Michele Chiuini
John Wiley & Sons, latest edition

Structure in Architecture: The Building of Buildings
Mario Salvadori with Robert Heller
Prentice-Hall, 1986

Understanding Structures
Fuller Moore
McGraw-Hill, 1999

Wood Design Manual and *CAN/CSA-086.1-94 and Commentary*
Canadian Wood Council

C. Other reference materials

Chen, Gang. *Building Construction: Project Management, Construction Administration, Drawings, Specs, Detailing Tips, Schedules, Checklists, and Secrets Others Don't Tell You (Architectural Practice Simplified, 2nd edition).* ArchiteG, Inc., A good introduction to the architectural practice and construction documents and service, including discussions of MasterSpec format and specification sections.

Chen, Gang. **LEED v4 Green Associate Exam Guide (LEED GA):** *Comprehensive Study Materials, Sample Questions, Mock Exam, Green Building LEED Certification, and Sustainability*, Book 2, LEED Exam Guide series, ArchiteG.com, the latest edition. ArchiteG, Inc. Latest Edition. This is a very comprehensive and concise book on the LEED Green Associate Exam. Some readers have passed the LEED Green Associate Exam by studying this book for 10 hours.

Ching, Francis. *Architecture: Form, Space, & Order.* Wiley, latest edition. It is one of the best architectural books that you can have. I still flip through it every now and then. It is a great book for inspiration.

Frampton, Kenneth. *Modern Architecture: A Critical History.* Thames and Hudson, London, latest edition. A valuable resource for architectural history.

Jarzombek, Mark M. (Author), Vikramaditya Prakash (Author), Francis D. K. Ching (Editor). *A Global History of Architecture.* Wiley, latest edition. A valuable and comprehensive resource for architectural history with 1000 b & w photos, 50 color photos, and 1500 b & w illustrations. It doesn't limit the topic on a Western perspective, but rather through a global vision.

Trachtenberg, Marvin and Isabelle Hyman. *Architecture: From Pre-history to Post-Modernism.* Prentice Hall, Englewood Cliffs, NJ latest edition. A valuable and comprehensive resource for architectural history.

D. Some Important Information about Architects and the Profession of Architecture

What Architects Do?

Architects plan and design houses, factories, office buildings, and other structures.

Duties

Architects typically do the following:

- meet with clients to determine objectives and requirements for structures
- give preliminary estimates on cost and construction time
- prepare structure specifications
- direct workers who prepare drawings and documents
- prepare scaled drawings, either with computer software or by hand
- prepare contract documents for building contractors
- manage construction contracts
- visit worksites to ensure that construction adheres to architectural plans
- seek new work by marketing and giving presentations

People need places to live, work, play, learn, shop, and eat. Architects are responsible for designing these places. They work on public or private projects and design both indoor and outdoor spaces. Architects can be commissioned to design anything from a single room to an entire complex of buildings.

Architects discuss the objectives, requirements, and budget of a project with clients. In some cases, architects provide various predesign services, such as feasibility and environmental impact studies, site selection, cost analyses, and design requirements.

Architects develop final construction plans after discussing and agreeing on the initial proposal with clients. These plans show the building's appearance and details of its construction. Accompanying these plans are drawings of the structural system; air-conditioning, heating, and ventilating systems; electrical systems; communications systems; and plumbing. Sometimes, landscape plans are included as well. In developing designs, architects must follow state and local building codes, zoning laws, fire regulations, and other ordinances, such as those requiring easy access to buildings for people who are disabled.

Computer-aided design and drafting (CADD) and building information modeling (BIM) have replaced traditional drafting paper and pencil as the most common methods for creating designs and construction drawings. However, hand-drawing skills are still required, especially during the conceptual stages of a project and when an architect is at a construction site.

As construction continues, architects may visit building sites to ensure that contractors follow the design, adhere to the schedule, use the specified materials, and meet work-quality standards. The job is not complete until all construction is finished, required tests are conducted, and construction costs are paid.

Architects may also help clients get construction bids, select contractors, and negotiate construction contracts.

Architects often collaborate with workers in related occupations, such as civil engineers, urban and regional planners, drafters, interior designers, and landscape architects.

Work Environment
Although architects usually work in an office, they must also travel to construction sites.

Architects held about 128,800 jobs in 2016. The largest employers of architects were as follows:

Architectural, engineering, and related services	68%
Self-employed workers	20%
Government	3%
Construction	2%

Architects spend much of their time in offices, where they meet with clients, develop reports and drawings, and work with other architects and engineers. They also visit construction sites to ensure clients' objectives are met and to review the progress of projects. Some architects work from home offices.

Work Schedules
Most architects work full time and many work additional hours, especially when facing deadlines. Self-employed architects may have more flexible work schedules.

How to Become an Architect
There are typically three main steps to becoming a licensed architect: completing a professional degree in architecture, gaining relevant experience through a paid internship, and passing the Architect Registration Examination.

Education
In all states, earning a professional degree in architecture is typically the first step to becoming an architect. Most architects earn their professional degree through a five-year Bachelor of Architecture degree program, intended for students with no previous architectural training. Many earn a master's degree in architecture, which can take one to five years in addition to the time spent earning a bachelor's degree. The amount of time required depends on the extent of the student's previous education and training in architecture.

A typical bachelor's degree program includes courses in architectural history and theory, building design with an emphasis on computer-aided design and drafting (CADD), structures, construction methods, professional practices, math, physical sciences, and liberal arts. Central to most architectural programs is the design studio, where students apply the skills and concepts learned in the classroom to create drawings and three-dimensional models of their designs.

Currently, thirty-four states require that architects hold a professional degree in architecture from one of the 123 schools of architecture accredited by the National Architectural Accrediting Board (NAAB). State licensing requirements can be found at the National Council of Architectural

Registration Boards (NCARB). In the states that do not have that requirement, applicants can become licensed with eight to thirteen years of related work experience in addition to a high school diploma. However, most architects in these states still obtain a professional degree in architecture.

Training

All state architectural registration boards require architecture graduates to complete a lengthy paid internship—generally three years of experience—before they may sit for the Architect Registration Examination. Most new graduates complete their training period by working at architectural firms through the Architectural Experience Program (AXP), a program run by NCARB that guides students through the internship process. Some states allow a portion of the training to occur in the offices of employers in related careers, such as engineers and general contractors. Architecture students who complete internships while still in school can count some of that time toward the three-year training period.

Interns in architectural firms may help design part of a project. They may help prepare architectural documents and drawings, build models, and prepare construction drawings on CADD. Interns may also research building codes and write specifications for building materials, installation criteria, the quality of finishes, and other related details. Licensed architects will take the documents that interns produce, make edits to them, finalize plans, and then sign and seal the documents.

Licenses, Certifications, and Registrations

All states and the District of Columbia require architects to be licensed. Licensing requirements typically include completing a professional degree in architecture, gaining relevant experience through a paid internship, and passing the Architect Registration Examination.

Most states also require some form of continuing education to keep a license, and some additional states are expected to adopt mandatory continuing education. Requirements vary by state but usually involve additional education through workshops, university classes, conferences, self-study courses, or other sources.

A growing number of architects voluntarily seek certification from NCARB. This certification makes it easier to become licensed in other states, because it is the primary requirement for reciprocity of licensing among state boards that are NCARB members. In 2014, approximately one-third of all licensed architects had the certification.

Advancement

After many years of work experience, some architects advance to become architectural and engineering managers. These managers typically coordinate the activities of employees and may work on larger construction projects.

Important Qualities

Analytical skills. Architects must understand the content of designs and the context in which they were created. For example, architects must understand the locations of mechanical systems and how those systems affect building operations.

Communication skills. Architects share their ideas, both in oral presentations and in writing, with clients, other architects, and workers who help prepare drawings. Many also give presentations to explain their ideas and designs.

Creativity. Architects design the overall look of houses, buildings, and other structures. Therefore, the final product should be attractive and functional.

Organizational skills. Architects often manage contracts. Therefore, they must keep records related to the details of a project, including total cost, materials used, and progress.

Technical skills. Architects need to use CADD technology to create plans as part of building information modeling (BIM).

Visualization skills. Architects must be able to see how the parts of a structure relate to each other. They also must be able to visualize how the overall building will look once completed.

Pay

The median annual wage for architects was $79,380 in May 2018. The median wage is the wage at which half the workers in an occupation earned more than that amount and half earned less. The lowest 10 percent earned less than $48,020, and the highest ten percent earned more than $138,120.

In May 2018, the median annual wages for architects in the top industries in which they worked were as follows:

Government	$92,940
Architectural, engineering, and related services	$78,460
Construction	$78,110

Most architects work full time and many work additional hours, especially when facing deadlines. Self-employed architects may have more flexible work hours.

Job Outlook

Employment of architects is projected to grow 4 percent from 2016 to 2026, slower than the average for all occupations.

Architects will be needed to make plans and designs for the construction and renovation of homes, offices, retail stores, and other structures. Many school districts and universities are expected to build new facilities or renovate existing ones. In addition, demand is expected for more healthcare facilities as the baby-boomer population ages and as more individuals use healthcare services. The construction of new retail establishments may also require more architects.

Demand is projected for architects with a knowledge of "green design," also called sustainable design. Sustainable design emphasizes the efficient use of resources, such as energy and water conservation; waste and pollution reduction; and environmentally friendly design, specifications, and materials. Rising energy costs and increased concern about the environment have led to many new buildings being built with more sustainable designs.

The use of CADD and, more recently, BIM, has made architects more productive. These technologies have allowed architects to do more work without the help of drafters while making it easier to share the work with engineers, contractors, and clients.

Job Prospects

With a high number of students graduating with degrees in architecture, very strong competition for internships and jobs is expected. Competition for jobs will be especially strong at the most prestigious architectural firms. Those with up-to-date technical skills—including a strong grasp of CADD and BIM—and experience in sustainable design will have an advantage.

Employment of architects is strongly tied to the activity of the construction industry. Therefore, these workers may experience periods of unemployment when there is a slowdown in requests for new projects or when the overall level of construction falls.

State & Area Data
Occupational Employment Statistics (OES)

The Occupational Employment Statistics (OES) program produces employment and wage estimates annually for over 800 occupations. These estimates are available for the nation as a whole, for individual states, and for metropolitan and nonmetropolitan areas. The link below goes to OES data maps for employment and wages by state and area.
htttps://www.bls.gov/oes/current/oes171011.htm#st

Projections Central

Occupational employment projections are developed for all states by Labor Market Information (LMI) or individual state Employment Projections offices. All state projections data are available at www.projectionscentral.com. Information on this site allows projected employment growth for an occupation to be compared among states or to be compared within one state. In addition, states may produce projections for areas; there are links to each state's websites where these data may be retrieved.

CareerOneStop

CareerOneStop includes hundreds of occupational profiles with data available by state and metro area. There are links in the left-hand side menu to compare occupational employment by state and occupational wages by local area or metro area. There is also a salary info tool to search for wages by zip code.

Related Occupations

Architects design buildings and related structures. Construction managers, like architects, also plan and coordinate activities concerned with the construction and maintenance of buildings and facilities. Others who engage in similar work are landscape architects, civil engineers, urban and regional planners, and designers, including interior designers, commercial and industrial designers, and graphic designers.

Sources of Additional Information

Disclaimer:
Links to non-BLS Internet sites are provided for your convenience and do not constitute an endorsement.

Information about education and careers in architecture can be obtained from:
- The American Institute of Architects, 1735 New York Ave. NW., Washington, DC 20006. Internet: http://www.aia.org
- National Architectural Accrediting Board: http://www.naab.org/
- National Council of Architectural Registration Boards, Suite 1100K, 1801 K St. NW., Washington, D.C. 20006. Internet: http://www.ncarb.org
  OOH ONET Codes 17-1011.00"

Source: Bureau of Labor Statistics, U.S. Department of Labor, *Occupational Outlook Handbook*, Architects, on the Internet at https://www.bls.gov/ooh/architecture-and-engineering/architects.htm (visited June 06, 2019).

Last Modified Date: Friday, April 12, 2019

Note:
Please check the website above for the latest information.

E. AIA Compensation Survey

Every 3 years, AIA publishes a Compensation Survey for various positions at architectural firms across the country. It is a good idea to find out the salary before you make the final decision to become an architect. If you are already an architect, it is also a good idea to determine if you are underpaid or overpaid.

See following link for some sample pages for the latest AIA Compensation Survey:

https://www.aia.org/resources/8066-aia-compensation-report

F. So ... You would Like to Study Architecture

To study architecture, you need to learn how to draft, how to understand and organize spaces and the interactions between interior and exterior spaces, how to do design, and how to communicate effectively. You also need to understand the history of architecture.

As an architect, a leader for a team of various design professionals, you not only need to know architecture, but also need to understand enough of your consultants' work to be able to coordinate them. Your consultants include soils and civil engineers, landscape architects, structural, electrical, mechanical, and plumbing engineers, interior designers, sign consultants, etc.

There are two major career paths for you in architecture: practice as an architect or teach in colleges or universities. The earlier you determine which path you are going to take, the more likely you will be successful at an early age. Some famous and well-respected architects, like my USC alumnus Frank Gehry, have combined the two paths successfully. They teach at the universities and have their own architectural practice. Even as a college or university professor, people respect you more if you have actual working experience and have some built projects. If you only teach in colleges or universities but have no actual working experience and have no built projects, people will consider you as a "paper" architect, and they are not likely to take you seriously, because they will think you probably do not know how to put a real building together.

In the U.S., if you want to practice architecture, you need to obtain an architect's license. It requires a combination of passing scores on the Architectural Registration Exam (ARE) and 8 years of education and/or qualified working experience, including at least 1 year of working experience in the U.S. Your working experience needs to be under the supervision of a licensed architect to be counted as qualified working experience for your architect's license.

If you work for a landscape architect or civil engineer or structural engineer, some states' architectural licensing boards will count your experience at a discounted rate for the qualification of your architect's license. For example, 2 years of experience working for a civil engineer may be counted as 1 year of qualified experience for your architect's license. You need to contact your state's architectural licensing board for specific licensing requirements for your state.

If you want to teach in colleges or universities, you probably want to obtain a master's degree or a Ph.D. It is not very common for people in the architectural field to have a Ph.D. One reason is that there are few Ph.D. programs for architecture. Another reason is that architecture is considered a profession and requires a license. Many people think an architect's license is more important than a Ph.D. degree. In many states, you need to have an architect's license to even use the title "architect," or the terms "architectural" or "architecture" to advertise your service. You cannot call yourself an architect if you do not have an architect's license, even if you have a Ph.D. in architecture. Violation of these rules brings punishment.

To become a tenured professor, you need to have a certain number of publications and pass the evaluation for the tenure position. Publications are very important for tenure track positions. Some people say for the tenured track positions in universities and colleges, it is "publish or perish."

The American Institute of Architects (AIA) is the national organization for the architectural profession. Membership is voluntary. There are different levels of AIA membership. Only licensed architects can be (full) AIA members. If you are an architectural student or an intern but not a licensed architect yet, you can join as an associate AIA member. Contact AIA for detailed information.

The National Council of Architectural Registration Boards (NCARB) is a nonprofit federation of architectural licensing boards. It has some very useful programs, such as IDP, to assist you in obtaining your architect's license. Contact NCARB for detailed information.

Back Page Promotion

You may be interested in some other books written by Gang Chen:

A. ARE Mock Exam series & ARE Exam Guide series. See the following link:
http://www.GreenExamEducation.com

B. LEED Exam Guides series. See the following link:
http://www.GreenExamEducation.com

C. *Building Construction: Project Management, Construction Administration, Drawings, Specs, Detailing Tips, Schedules, Checklists, and Secrets Others Don't Tell You (Architectural Practice Simplified, 2nd edition)*
http://www.GreenExamEducation.com

D. *Planting Design Illustrated*
http://www.GreenExamEducation.com

ARE Mock Exam Series & ARE Exam Guide Series

ARE 5.0 Mock Exam Series

Practice Management (PcM) ARE 5.0 Mock Exam (Architect Registration Examination): ARE 5.0 Overview, Exam Prep Tips, Hotspots, Case Studies, Drag-and-Place, Solutions and Explanations. **ISBN**: 9781612650388

Project Management (PjM) ARE 5.0 Mock Exam (Architect Registration Examination): ARE 5.0 Overview, Exam Prep Tips, Hotspots, Case Studies, Drag-and-Place, Solutions and Explanations. **ISBN**: 9781612650371

Programming & Analysis (PA) ARE 5.0 Mock Exam (Architect Registration Exam): ARE 5.0 Overview, Exam Prep Tips, Hotspots, Case Studies, Drag-and-Place, Solutions and Explanations. **ISBN**: 9781612650326

Project Planning & Design (PPD) ARE 5.0 Mock Exam (Architect Registration Examination): ARE 5.0 Overview, Exam Prep Tips, Hotspots, Case Studies, Drag-and-Place, Solutions and Explanations. **ISBN**: 9781612650296

Project Development & Documentation (PDD) ARE 5.0 Mock Exam (Architect Registration Examination): ARE 5.0 Overview, Exam Prep Tips, Hotspots, Case Studies, Drag-and-Place, Solutions and Explanations
ISBN: 9781612650258

Construction & Evaluation (CE) ARE 5.0 Mock Exam (Architect Registration Examination): ARE 5.0 Overview, Exam Prep Tips, Hotspots, Case Studies, Drag-and-Place, Solutions and Explanations
ISBN: 9781612650241

Mock California Supplemental Exam (CSE of Architect Registration Examination): CSE Overview, Exam Prep Tips, General Section and Project Scenario Section, Questions, Solutions and Explanations. **ISBN**: 9781612650159

ARE 5.0 Exam Guide Series

Practice Management (PcM) ARE 5.0 Exam Guide (Architect Registration Examination): ARE 5.0 Overview, Exam Prep Tips, Guide, and Critical Content. **ISBN**: 9781612650333

Project Management (PjM) ARE 5.0 Exam Guide (Architect Registration Examination): ARE 5.0 Overview, Exam Prep Tips, Guide, and Critical Content. **ISBN**: 9781612650418

Programming & Analysis (PA) ARE 5.0 Exam Guide (Architect Registration Examination): ARE 5.0 Overview, Exam Prep Tips, Guide, and Critical Content.
ISBN: 9781612650487

Construction and Evaluation (CE) ARE 5 Exam Guide (Architect Registration Exam):
ARE 5.0 Overview, Exam Prep Tips, Guide, and Critical Content
ISBN: 9781612650432

Other books in the ARE 5.0 Exam Guide Series are being produced. Our goal is to produce one mock exam book *plus* one guidebook for each of the ARE 5.0 exam divisions. See the following link for the latest information:
http://www.GreenExamEducation.com

LEED Exam Guides series: Comprehensive Study Materials, Sample Questions, Mock Exam, Building LEED Certification and Going Green

LEED (Leadership in Energy and Environmental Design) is the most important trend of development, and it is revolutionizing the construction industry. It has gained tremendous momentum and has a profound impact on our environment.

From LEED Exam Guides series, you will learn how to

1. Pass the LEED Green Associate Exam and various LEED AP + exams (each book will help you with a specific LEED exam).

2. Register and certify a building for LEED certification.

3. Understand the intent for each LEED prerequisite and credit.

4. Calculate points for a LEED credit.

5. Identify the responsible party for each prerequisite and credit.

6. Earn extra credit (exemplary performance) for LEED.

7. Implement the local codes and building standards for prerequisites and credit.

8. Receive points for categories not yet clearly defined by USGBC.

There is currently NO official book on the LEED Green Associate Exam, and most of the existing books on LEED and LEED AP are too expensive and too complicated to be practical and helpful. The pocket guides in LEED Exam Guides series fill in the blanks, demystify LEED, and uncover the tips, codes, and jargon for LEED as well as the true meaning of "going green." They will set up a solid foundation and fundamental framework of LEED for you. Each book in the LEED Exam Guides series covers every aspect of one or more specific LEED rating system(s) in plain and concise language and makes this information understandable to all people.

These pocket guides are small and easy to carry around. You can read them whenever you have a few extra minutes. They are indispensable books for all people—administrators; developers; contractors; architects; landscape architects; civil, mechanical, electrical, and plumbing engineers; interns; drafters; designers; and other design professionals.

Why is the LEED Exam Guides series needed?

A number of books are available that you can use to prepare for the LEED exams:

1. *USGBC Reference Guides.* You need to select the correct version of the *Reference Guide* for your exam.

   The *USGBC Reference Guides* are comprehensive, but they give too much information. For example, *The LEED 2009 Reference Guide for Green Building Design and Construction (BD&C)* has about 700 oversized pages. Many of the calculations in the books are too detailed for the exam. They are also expensive (approximately $200 each, so most people may not buy them for their personal use, but instead, will seek to share an office copy).

   It is good to read a reference guide from cover to cover if you have the time. The problem is not too many people have time to read the whole reference guide. Even if you do read the whole guide, you may not remember the important issues to pass the LEED exam. You need to reread the material several times before you can remember much of it.

   Reading the reference guide from cover to cover without a guidebook is a difficult and inefficient way of preparing for the LEED AP Exam, because you do NOT know what USGBC and GBCI are looking for in the exam.

2. The USGBC workshops and related handouts are concise, but they do not cover extra credits (exemplary performance). The workshops are expensive, costing approximately $450 each.

3. Various books published by a third party are available on Amazon, bn.com and books.google.com. However, most of them are not very helpful.

   There are many books on LEED, but not all are useful.

   LEED Exam Guides series will fill in the blanks and become a valuable, reliable source:

   a. They will give you more information for your money. Each of the books in the LEED Exam Guides series has more information than the related USGBC workshops.

   b. They are exam-oriented and more effective than the USGBC reference guides.

   c. They are better than most, if not all, of the other third-party books. They give you comprehensive study materials, sample questions and answers, mock exams and answers, and critical information on building LEED certification and going green. Other third-party books only give you a fraction of the information.

   d. They are comprehensive yet concise. They are small and easy to carry around. You can read them whenever you have a few extra minutes.

   e. They are great timesavers. I have highlighted the important information that you need to understand and MEMORIZE. I also make some acronyms and short sentences to help you easily remember the credit names.

It should take you about 1 or 2 weeks of full-time study to pass each of the LEED exams. I have met people who have spent 40 hours to study and passed the exams.

You can find sample texts and other information on the LEED Exam Guides series in customer discussion sections under each of my book's listing on Amazon, bn.com and books.google.com.

What others are saying about *LEED GA Exam Guide* (Book 2, LEED Exam Guide series):

"Finally! A comprehensive study tool for LEED GA Prep!

"I took the 1-day Green LEED GA course and walked away with a power point binder printed in very small print—which was missing MUCH of the required information (although I didn't know it at the time). I studied my little heart out and took the test, only to fail it by 1 point. Turns out I did NOT study all the material I needed to in order to pass the test. I found this book, read it, marked it up, retook the test, and passed it with a 95%. Look, we all know the LEED GA exam is new and the resources for study are VERY limited. This one is the VERY best out there right now. I highly recommend it."
—**ConsultantVA**

"Complete overview for the LEED GA exam

"I studied this book for about 3 days and passed the exam … if you are truly interested in learning about the LEED system and green building design, this is a great place to start."
—**K.A. Evans**

"A Wonderful Guide for the LEED GA Exam

"After deciding to take the LEED Green Associate exam, I started to look for the best possible study materials and resources. From what I thought would be a relatively easy task, it turned into a tedious endeavor. I realized that there are vast amounts of third-party guides and handbooks. Since the official sites offer little to no help, it became clear to me that my best chance to succeed and pass this exam would be to find the most comprehensive study guide that would not only teach me the topics, but would also give me a great background and understanding of what LEED actually is. Once I stumbled upon Mr. Chen's book, all my needs were answered. This is a great study guide that will give the reader the most complete view of the LEED exam and all that it entails.

"The book is written in an easy-to-understand language and brings up great examples, tying the material to the real world. The information is presented in a coherent and logical way, which optimizes the learning process and does not go into details that will not be needed for the LEED Green Associate Exam, as many other guides do. This book stays dead on topic and keeps the reader interested in the material.

"I highly recommend this book to anyone that is considering the LEED Green Associate Exam. I learned a great deal from this guide, and I am feeling very confident about my chances for passing my upcoming exam."
—**Pavel Geystrin**

"Easy to read, easy to understand

"I have read through the book once and found it to be the perfect study guide for me. The author does a great job of helping you get into the right frame of mind for the content of the exam. I had started by studying the Green Building Design and Construction reference guide for LEED projects produced by the USGBC. That was the wrong approach, simply too much information with very little retention. At 636 pages in textbook format, it would have been a daunting task to get through it. Gang Chen breaks down the points, helping to minimize the amount of information but maximizing the content I was able to absorb. I plan on going through the book a few more times, and I now believe I have the right information to pass the LEED Green Associate Exam."
—**Brian Hochstein**

"All in one—LEED GA prep material

"Since the LEED Green Associate exam is a newer addition by USGBC, there is not much information regarding study material for this exam. When I started looking around for material, I got really confused about what material I should buy. This LEED GA guide by Gang Chen is an answer to all my worries! It is a very precise book with lots of information, like how to approach the exam, what to study and what to skip, links to online material, and tips and tricks for passing the exam. It is like the 'one stop shop' for the LEED Green Associate Exam. I think this book can also be a good reference guide for green building professionals. A must-have!"
—**SwatiD**

"An ESSENTIAL LEED GA Exam Reference Guide

"This book is an invaluable tool in preparation for the LEED Green Associate (GA) Exam. As a practicing professional in the consulting realm, I found this book to be all-inclusive of the preparatory material needed for sitting the exam. The information provides clarity to the fundamental and advanced concepts of what LEED aims to achieve. A tremendous benefit is the connectivity of the concepts with real-world applications.

"The author, Gang Chen, provides a vast amount of knowledge in a very clear, concise, and logical media. For those that have not picked up a textbook in a while, it is very manageable to extract the needed information from this book. If you are taking the exam, do yourself a favor and purchase a copy of this great guide. Applicable fields: Civil Engineering, Architectural Design, MEP, and General Land Development."
—**Edwin L. Tamang**

Note:
*Other books in the **LEED Exam Guides series** are published or in the process of being produced. At least **one book will eventually be produced for each of the LEED exams**. The series include:*

LEED v4 Green Associate Exam Guide (LEED GA): *Comprehensive Study Materials, Sample Questions, Mock Exam, Green Building LEED Certification, and Sustainability,* LEED Exam Guide series, ArchiteG.com. Latest Edition.

***LEED GA MOCK EXAMS (LEED v4):** Questions, Answers, and Explanations: A Must-Have for the LEED Green Associate Exam, Green Building LEED Certification, and Sustainability*, LEED Exam Guide series, ArchiteG.com. Latest Edition

***LEED v4 BD&C EXAM GUIDE:** A Must-Have for the LEED AP BD+C Exam: Comprehensive Study Materials, Sample Questions, Mock Exam, Green Building Design and Construction, LEED Certification, and Sustainability*, LEED Exam Guide series, ArchiteG.com. Latest Edition.

***LEED v4 BD&C MOCK EXAMS:** Questions, Answers, and Explanations: A Must-Have for the LEED AP BD+C Exam, Green Building LEED Certification, and Sustainability*, LEED Exam Guide series, ArchiteG.com. Latest Edition.

***LEED v4 ID&C Exam Guide:** A Must-Have for the LEED AP ID+C Exam: Study Materials, Sample Questions, Green Interior Design and Construction, Green Building LEED Certification, and Sustainability*, LEED Exam Guide series, ArchiteG.com. Latest Edition.

***LEED v4 AP ID+C MOCK EXAM:** Questions, Answers, and Explanations: A Must-Have for the LEED AP ID+C Exam, Green Building LEED Certification, and Sustainability.* LEED Exam Guide series, ArchiteG.com. Latest Edition.

***LEED v4 AP O+M MOCK EXAM: Questions, Answers, and Explanations:** A Must-Have for the LEED AP O+M Exam, Green Building LEED Certification, and Sustainability.* LEED Exam Guide series, ArchiteG.com. Latest Edition.

***LEED v4 O&M EXAM GUIDE:** A Must-Have for the LEED AP O+M Exam: Comprehensive Study Materials, Sample Questions, Mock Exam, Green Building Operations and Maintenance, LEED Certification, and Sustainability*, LEED Exam Guide series, ArchiteG.com. Latest Edition.

***LEED v4 HOMES EXAM GUIDE:** A Must-Have for the LEED AP Homes Exam: Comprehensive Study Materials, Sample Questions, Mock Exam, Green Building LEED Certification, and Sustainability*, LEED Exam Guide series, ArchiteG.com. Latest Edition.

***LEED v4 ND EXAM GUIDE:** A Must-Have for the LEED AP Neighborhood Development Exam: Comprehensive Study Materials, Sample Questions, Mock Exam, Green Building LEED Certification, and Sustainability*, LEED Exam Guide series, ArchiteG.com. Latest Edition.

How to order these books:
You can order the books listed above at:
http://www.GreenExamEducation.com

OR
http://www.ArchiteG.com

Building Construction

Project Management, Construction Administration, Drawings, Specs, Detailing Tips, Schedules, Checklists, and Secrets Others Don't Tell You (Architectural Practice Simplified, 2nd edition)

Learn the Tips, Become One of Those Who Know Building Construction and Architectural Practice, and Thrive!

For architectural practice and building design and construction industry, there are two kinds of people: those who know, and those who don't. The tips of building design and construction and project management have been undercover—until now.

Most of the existing books on building construction and architectural practice are too expensive, too complicated, and too long to be practical and helpful. This book simplifies the process to make it easier to understand and uncovers the tips of building design and construction and project management. It sets up a solid foundation and fundamental framework for this field. It covers every aspect of building construction and architectural practice in plain and concise language and introduces it to all people. Through practical case studies, it demonstrates the efficient and proper ways to handle various issues and problems in architectural practice and building design and construction industry.

It is for ordinary people and aspiring young architects as well as seasoned professionals in the construction industry. For ordinary people, it uncovers the tips of building construction; for aspiring architects, it works as a construction industry survival guide and a guidebook to shorten the process in mastering architectural practice and climbing up the professional ladder; for seasoned architects, it has many checklists to refresh their memory. It is an indispensable reference book for ordinary people, architectural students, interns, drafters, designers, seasoned architects, engineers, construction administrators, superintendents, construction managers, contractors, and developers.

You will learn:
1. How to develop your business and work with your client.
2. The entire process of building design and construction, including programming, entitlement, schematic design, design development, construction documents, bidding, and construction administration.
3. How to coordinate with governing agencies, including a county's health department and a city's planning, building, fire, public works departments, etc.
4. How to coordinate with your consultants, including soils, civil, structural, electrical, mechanical, plumbing engineers, landscape architects, etc.
5. How to create and use your own checklists to do quality control of your construction documents.
6. How to use various logs (i.e., RFI log, submittal log, field visit log, etc.) and lists (contact list, document control list, distribution list, etc.) to organize and simplify your work.
7. How to respond to RFI, issue CCDs, review change orders, submittals, etc.
8. How to make your architectural practice a profitable and successful business.

Planting Design Illustrated
A Must-Have for Landscape Architecture: A Holistic Garden Design Guide with Architectural and Horticultural Insight, and Ideas from Famous Gardens in Major Civilizations

One of the most significant books on landscaping!

This is one of the most comprehensive books on planting design. It fills in the blanks of the field and introduces poetry, painting, and symbolism into planting design. It covers in detail the two major systems of planting design: formal planting design and naturalistic planting design. It has numerous line drawings and photos to illustrate the planting design concepts and principles. Through in-depth discussions of historical precedents and practical case studies, it uncovers the fundamental design principles and concepts, as well as the underpinning philosophy for planting design. It is an indispensable reference book for landscape architecture students, designers, architects, urban planners, and ordinary garden lovers.

What Others Are Saying about *Planting Design Illustrated* ...

"I found this book to be absolutely fascinating. You will need to concentrate while reading it, but the effort will be well worth your time."
—**Bobbie Schwartz, former president of APLD (Association of Professional Landscape Designers) and author of** *The Design Puzzle: Putting the Pieces Together.*

"This is a book that you have to read, and it is more than well worth your time. Gang Chen takes you well beyond what you will learn in other books about basic principles like color, texture, and mass."
—**Jane Berger, editor & publisher of gardendesignonline**

"As a longtime consumer of gardening books, I am impressed with Gang Chen's inclusion of new information on planting design theory for Chinese and Japanese gardens. Many gardening books discuss the beauty of Japanese gardens, and a few discuss the unique charms of Chinese gardens, but this one explains how Japanese and Chinese history, as well as geography and artistic traditions, bear on the development of each country's style. The material on traditional Western garden planting is thorough and inspiring, too. *Planting Design Illustrated* definitely rewards repeated reading and study. Any garden designer will read it with profit."
—**Jan Whitner, editor of the** *Washington Park Arboretum Bulletin*

"Enhanced with an annotated bibliography and informative appendices, *Planting Design Illustrated* offers an especially "reader friendly" and practical guide that makes it a very strongly recommended addition to personal, professional, academic, and community library gardening & landscaping reference collection and supplemental reading list."
—**Midwest Book Review**

"Where to start? *Planting Design Illustrated* is, above all, fascinating and refreshing! Not something the lay reader encounters every day, the book presents an unlikely topic in an easily digestible, easy-to-follow way. It is superbly organized with a comprehensive table of contents, bibliography, and appendices. The writing, though expertly informative, maintains its accessibility throughout and is a joy to read. The detailed and beautiful illustrations expanding on the concepts presented were my favorite portion. One of the finest books I've encountered in this contest in the past 5 years."
—**Writer's Digest 16th Annual International Self-Published Book Awards Judge's Commentary**

"The work in my view has incredible application to planting design generally and a system approach to what is a very difficult subject to teach, at least in my experience. Also featured is a very beautiful philosophy of garden design principles bordering poetry. It's my strong conviction that this work needs to see the light of day by being published for the use of professionals, students & garden enthusiasts."
—**Donald C. Brinkerhoff, FASLA, chairman and CEO of Lifescapes International, Inc.**

Index

3016 rule, 14, 30, 32
A/E, 20
accrual-basis method, 78, 79
Agency, 82, 94
AIA, 15, 112, 114
Architect's Handbook of Professional Practice, 37, 61, 62, 63, 74, 98
Architectural Experience Program, 13, 17, 19, 21, 22, 53, 78
ARE Guidelines, 18
ARE Mock Exam, 3, 9, 13, 15, 29, 41, 61, 115, 116, 117
ARE® Guidelines, 19
AXP, 13, 17, 19, 20, 21, 22, 45, 53, 66, 78
AXP Portfolio, 20
B101, Standard Form of Agreement Between Owner and Architect, 66, 77, 81
B141, *Standard Form of Agreement between Owner and Architect with Standard Form of Architect's Services*, 66
BD&C, 85
Best Practices, 76
breaks, 3, 14, 33, 122
Business interruption insurance, 61
C corporation, 43, 64
case studies, 9, 11, 17, 124, 125
cash-basis method, 78, 79
CE, 14, 17, 25, 98, 99, 100, 101
check, 19
check-all-that-apply, 11, 17
CMAR, 52, 76
Code of Ethics and Professional Conduct, 45, 49, 56, 65, 66, 72, 81, 99
Codes and standards, 14, 34
ConsensusDOCS, 50, 74
construction management at risk, 52, 76
Construction manager-agent, 63
construction manager-constructor, 43, 63
Cost-plus fee, 79
CSI, 37
CSI divisions, 39

design charrette, 48, 71
Design-assist contracting, 50, 73
Design-bid-build, 63, 76
Design-build, 63
double-entry method, 78
drag-and-place, 11, 17
Employment practice liability insurance, 61
exam content, 23, 28, 35
Exam Format & Time, 24
Fast-track, 76
Fiduciary duty, 82
field visit, 57, 124
General partnership, 63
Health insurance, 61
Hotspots, 11, 17
Hourly not to exceed, 79
ID&C, 85
IDP, 13, 17, 19, 20, 21, 26, 108, 114
Indemnity, 82
Index, 128
integrated project delivery, 47, 52, 53, 69, 76, 78
intern, 21
joint venture, 48, 53, 56, 71, 78, 81
LEED, 15, 115, 119, 120, 121, 122, 123
LEED for Health Care, 85
LEED for Retails, 84, 85
LEED-EB, 85
LEED-ND, 85
limited liability company, 48, 71
limited liability partnership, 43, 63
LLP, 37, 63, 99
MacLeamy curve, 53, 77
mnemonics, 14, 30, 31, 32
Mnemonics, 31, 32, 38, 39, 87
multiple choice, 11, 17
NCARB, 28
Negligence, 69
net multiplier, 46, 69
net revenue, 46, 69, 79

Note, 26, 36, 39, 61, 111, 122
O&M, 85
outside freelance consultant, 56, 81
overhead rate, 54, 79
PA, 14, 17, 98, 99, 100, 101
passing or failing percentage, 24
PcM, 14, 17, 98, 99, 100, 101
PDD, 14, 17, 28, 98, 99, 100, 101
Percentage of construction cost, 79
performance bond, 42, 47, 61, 70
physical exercise, 14, 33
PjM, 14, 17, 98, 99, 100, 101
PPD, 14, 15, 17, 35, 36, 86, 98, 99, 100, 101, 116
prime contract agreement, 56, 81
privity, 58, 82
profit and loss statement, 54, 68, 79
quality control, 55, 58, 72, 80, 83, 124
quantitative fill-in-the-blank, 11, 17
quick ratio, 47, 69
register, 23, 24

reporting hours, 19, 20
RFP, 43, 44, 62, 65
rolling clock, 23
routine, 14, 33
Rules of Conduct, 21
S corporation, 43, 44, 63, 64
scores, 24
section, 37
single studio, 58, 83
Six-Month Rule, 20
Sole proprietorship, 63, 71
Square-feet cost, 79
Stress, 102
teaming agreement, 53, 56, 78, 81
test-taking tips, 25
Tips, 115, 124
tort, 47, 69
U/A, 20
utilization ratio, 47, 69
vicarious liability, 51, 75

Made in the USA
Middletown, DE
20 January 2022